NUTRITION AND NUTRITIONAL PHYSIOLOGY OF THE FOX

A HISTORICAL PERSPECTIVE

WILLIAM L. LEOSCHKE, Ph.D.

Order this book online at www.trafford.com
or email orders@trafford.com

Most Trafford titles are also available at major online book retailers.

Printed in the United States of America.

ISBN: 978-1-4251-5101-0 (sc)
ISBN: 978-1-4251-5103-4 (e)

Trafford rev. 04/25/2011

www.trafford.com

North America & international
toll-free: 1 888 232 4444 (USA & Canada)
phone: 250 383 6864 • fax: 812 355 4082

Nutrition and Nutritional Physiology of the Fox

A Historical Perspective

William L. Leoschke, Ph.D.

Nutrition and Nutritional Physiology of the Fox
An Historical Perspective

William L. Leoschke, Ph.D.
Director of Fur Animal Nutrition Research
National Fur Foods Company
New Holstein, Wisconsin 53061
Emeritus Professor of Chemistry
Valparaiso University
Valparaiso, Indiana 46383-9615

Edited by Kathleen Mullen, Ph.D.
Professor of English Emerita
Valparaiso University

Book Design by Robert Sirko, MFA
Associate Professor of Graphic Design
Valparaiso University

Foreword

Nutrition and Nutritional Physiology Of The Fox – A Historical Perspective is written by Dr. William L. Leoschke, a fur animal nutrition scientist who is recognized worldwide for his basic research studies on mink and chinchilla nutrition at the University of Wisconsin, 1950-1959, and his numerous articles on mink nutrition in fur animal trade journals over a period of almost three decades, including the years 1960-1965 in association with Dr. R. Shackelford, geneticist, and Dr. G. Hartsought, veterinarian, in a monthly column in the *American Fur Breeder* entitled "Ask The Experts -- About Diseases -- About Genetics -- About Nutrition." This fur animal education program was followed by a similar column entitled "Dr. Leoschke On Mink Nutrition" from 1965-1973 in the *American Fur Breeder* and the *U.S. Fur Rancher*. In addition each year of "The Blue Book of Fur Farming" from 1973 to 1988 included a significant mink nutrition article by Dr. Leoschke.

Relative to the worldwide fur industry, Dr. Leoschke provided articles for the *British Fur Farmers Gazette* from 1966-1967 as well as a "Question and Answer" column. He has served as a member of both the 1968 and 1982 National Academy of Science National Research Committee that prepared the publication on "Nutrient Requirements of Mink and Foxes." In addition, Dr. Leoschke has provided scientific presentations at the International Congresses In Fur Animal Production initially in 1976 in Finland and every four years following: 1980, Denmark; 1984, France; 1988, Canada; 1992, Norway; 1996, Poland; 2000, Greece; and 2004, The Netherlands.

In addition to these scientific activities, Dr. Leoschke has been a consultant to National Fur Foods Company, New Holstein, Wisconsin, as Director of Fur Animal Nutrition Research since 1955. The basic goal of the experimental studies was the development of top performance pellet formulations for mink, fox, ferrets, and even opossums. After many years of intensive and extensive research, the goal has been achieved wherein the performance of mink and fox on commercial pellets has equaled that of common fresh/frozen/fortified cereal programs in terms of reproduction, lactation, and kit growth. Relative to fur development it has provided unique qualities of light leather, silkiness and a sharp color unequalled on any practical fresh/frozen/fortified cereal program within the worldwide fur industry.

The book is documented with more than 300 references and includes multiple tables. The scientific observations are presented in a manner useful for fox nutrition scientists, commercial fox cereal and pellet manufacturers, veterinarians, and practical fox ranchers throughout the world.

Preface

It appears that a very significant portion of my life has been associated with fur bearing animals, from keeping a rabbit colony as a teenager in Western New York to graduate studies at the University of Wisconsin in Madison, Wisconsin, involving mink and chinchilla nutritional studies to a half-century of fur animal nutrition research studies involving mink, fox and ferrets as a consultant to National Fur Foods, New Holstein, Wisconsin.

This book is dedicated to the many individuals who have conducted experimental studies with foxes over the period of the last century and especially to Hans Rimeslatten, Agricultural College of Norway, the father of fox nutrition research within the world fur production community.

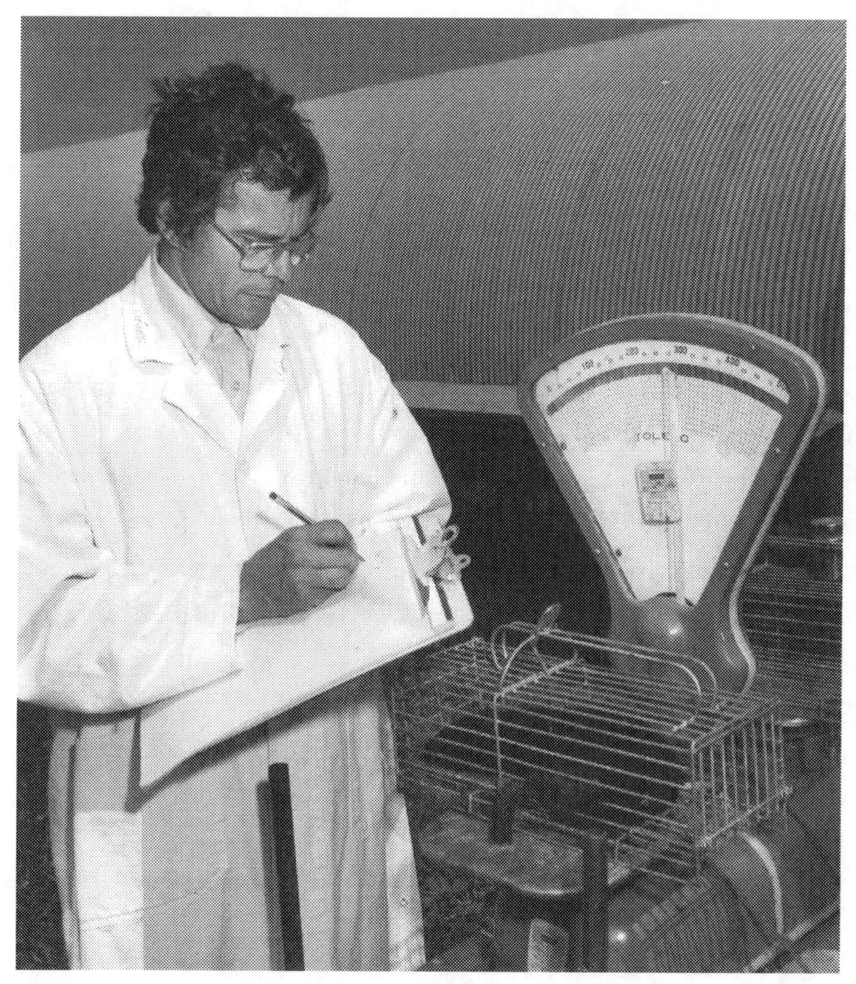

William L. Leoschke,

Chapter 1

Natural Diet of the Fox

Chapter 2

Fox Nutrition History—North America

Chapter 3

Anatomy and Physiology

Chapter 4

Nutrient Requirements

Chapter 5

Disorders Related to Nutrition Management

Chapter 6

Modern Fox Nutrition Management

Chapter 7

Practical Fox Ranch Diets-20th Century

Chapter 8

Modern Fox Nutrition-21st Century-Pellets

1

Natural Diet of the Fox

The family of the foxes (Canidae) belongs to the Vulpavines, the New World carnivores, which is also known as the "dog branch" of the carnivore family. The "cat branch," Viverravines, includes among others, the domestic cat. These branches have evolved separately for about 60 million years. During this evolution since the Paleocene Miacids, carnivores have adapted to diets low in carbohydrates (MacDonald, 1992). The Vulpavines are distinct from the Viverravines inasmuch as the "dog branch" are omnivores as adults while the "cat branch" remain carnivores as adults (Buddington et al., 1991).

Sub-groups of the fox family include:

Artic Fox	*(Alopex lag opus innuitus)*
Blue Fox	*(Alopex lag opus)*
Red Fox	*(Vulpes vulpes)*
Silver Fox	*(Vulpes vulpes fulva)*

Field Studies

In their natural environment, blue, red, and silver fox are opportunistic crepuscular predators and scavengers with a catholic diet as seen in Table 1.1 (Nelson, 1933; Hatfield, 1933; Chaddock, 1933; Errington, 1935; Hamilton, 1935; McGregor, 1942; Eadie, 1943; Hamilton and Cook, 1944; Heit, 1944; Glover, 1949; Schuler, 1951; Dodds, 1955; Cutter, 1958; Karpuleon, 1958; Korschgen, 1959; Chesemore, 1968; Kaikusalo, 1971; Goszcynski, 1974; Stoddard, 1974; Hewson and Kolb, 1975; Forbes and Lance, 1976; Watson, 1976; Richards, 1977; Frank, 1979; Raymond, 1980; Johnson, 1980; Jaksic, Schlatter, and Yanez, 1980; Tolonen, 1982; Dekker, 1983; Robertson and Whelan, 1987; Rouvinen, 1987; Pegageorgious et al., 1988).

TABLE 1.1 Natural Diet of the Fox

A. Large Mammals
Hedgehog, muskrat, opossum, porcupine, rabbits (cottontail, hares, jackrabbit and snowshoe), raccoon, roe deer, skunks, squirrels (flying, fox, grey and red), and woodchucks.

B. Smaller Mammals
Chipmunk, pocket gopher, lemmings, mice (deer, dormice, field, Flemming, grasshopper, harvest, jumping, meadow, pine, pocket, red backed, Western harvest, white-footed and wood), moles (hairy tailed, meadow, and star-nosed), rats (cotton, kangaroo, and Norway), shrew* (least, long tailed, and short tailed), voles (bank, common, field, red-backed, and water) and weasel.

C. Game Birds and Their Eggs
Ducks (mallard, sea, and shelduck), goose (edie), grouse (red, ruffed wary), gulls, owl (short-eared), partridge, passerines, pheasants (ring-necked), Ptarmigan, quail, and wild turkey.

D. Small Non-game Birds and Their Eggs

Blackbird, blue jay, bunting (lark), chickadee, cowbird, crow, flicker, hawks, junco, larks (horned), robin, sparrows (common, field, and slate-colored), and thrush.

E. Reptiles

Lizard (Eastern fence), roadrunner (six-liner), snakes (dekays, garter, and Stoddard), toad (Texas horned) and turtles.

F. Misc.

Fish**, frogs, and sand***

G. Carrion

Badger, caribou, cattle, deer (red and white-tailed), horse, poultry, sheep, swine, and walrus.

H. Fruits and Seeds

Apples (common and thorn), arrow wood, beechnuts, birch, berries (blue, black, dew, elder, huckle, rasp-, and straw-), cherries (choke, fire, wild, and wild black), grapes, hawthorn, hazel, juniper, maple leaf, oak acorns, persimmon seeds, wild raisin, and wild sarsaparilla.

I. Plants and Shrubs

Brambles, grasses, ragweed, shadbush, woodland, and fence row shrub.

---

* Generally, fox reject shrews because of low palatability due to secretions of flank gland (Richards, 1977).

** In its natural surroundings, in sharp contrast to the mink, fox eat fish only occasionally (Linscombe et al., 1982; Samuel and Nelson, 1982; Rouvinen, 1987).

*** Sand, leaves, sticks may be consumed to alleviate the irritation of the stomach mucosa by parasites which produce a toxin which is extremely painful to the stomach wall.

The winter diet of the fox is primarily mammals while the summer diet is based more on fruit. The arctic fox on St. Lawrence Island, Bering Sea, actually provide cached supplies for winter food when the birds are absent (Chokebore, 1968; Fay and Stephenson, 1989). Unlike the mink, foxes chop up their food and swallow it in chunks, as noted by Dixon (1925) who reported that the largest meal for a gray fox was 80 grams, a brush rabbit, and for a kit fox, 57 grams, a jack rabbit.

2

Fox Nutrition History

North America

In terms of the world history of fur as human apparel, North America has played a major role. The Chinese have used fur from about 500 B.C. and fur garments are recorded as being used in both ancient Greece and Rome. In medieval times, the use of fur was strictly controlled and only royalty, nobility, and certain clerics and layman were allowed to wear fur. Ermine could be used only for royal robes. In the middle ages and in Chaucer's time (1400 AD) beaver was used for making headdresses and was the only fur worn by the common people. The English custom of using ermine as a trimming for judges' robes has survived from the middle ages to this day. Furs became available to the general population for the first time following the discovery of North America inasmuch as the limited supply in Northern Europe restricted their use to the wealthy. The Hudson's Bay Company was founded in 1670 and their traders spread across the continent of North America in search of furs.

Fur animals are part of the American heritage. Exploration of the Northeastern part of the continent was initiated by French fur traders. The fur industry in terms of beaver, chinchilla, fox, muskrat, and mink is still an important part of the American scene. The beginning of commercial fur farming in North America took place in the early 1860s when a pair of young silver foxes were dug out of a den on Prince Edward Island.

Practical Fox Ranch Diets

TABLE 2.2 An Early Yearly Feeding Schedule*

Ranch Year	Breeding 12/15-3/15	Whelping to 6/1	Growth to 9/15	Fur to 11/1**	Finishing to Pelting
Ingredients					
Cereals					
Fortified Cereal	20		35		
Fortified Cereal w/Meat	40		40	50	
Wheat Germ		5			
Bran		5		3	5
Beet Pulp					3
Vitamins	2		2	2	
Liver	5	10	5	5	
Horsemeat	55	20	40	15***	10
Tripe/Spleen	10	15	10	25****	20
Green Bone	3		5	5	5
Vegetables/Figs	5	5	5	5	5

* Avery (1947), **Pups kept lean until 11/15 for higher feed intake in the final weeks of fur development.

*** Reduce horsemeat for sharper color in pelts marketed.

**** Lower fat levels via trimmed tripe for higher protein intake.

Note: Avery was a very observant fox rancher, noting loss of sharp color in fox pelts with high levels of horsemeat in the fox diet. Horsemeat, especially rancid horsemeat with peroxidized polyunsaturated fatty acids, can yield "fire engine red" pelts in dark mink.

Fox Nutrition Research—World Wide

Research on the nutritional requirements of fur animals was initiated in the United States at Cornell University, the University of Wisconsin, and Oregon State University in 1946.

In 1939 the Agricultural University of Norway included foxes in its research program. In 1947, Sweden initiated fur animal nutrition research at the State Institute of Animal Research at Uppsala. The Danish Fur Breeders Association established a silver fox research farm which was later donated to the government. In 1963, The Finnish Fur Breeders Association established the Yrjo Helvels foundation in memory of its earlier director and furnished a fur animal research farm (Anon, 1997).

3

Anatomy and Physiology

Anatomy

Body Composition

Adipose Tissue

TABLE 3.1 Fatty Acid Composition—Subcutaneous Adipose Tissue

Carbon-#	Name	Arctic*	Blue Fox*		Silver Fox**	
			Animal	Fish	Animal	Fish
14:0	Myristic	4.2				
14:1	Myristoleic	0.9				
16:0	Palmitic	2.9				
16:1	Palmitoleic	41.2				
17:0	Heptadecanoic	0.3				
18:0	Stearic	14.3				
18:1	Oleic	2.9				
18:2w6	Linoleic	18.5	7.0	4.8	7.7	6.1
18:3w3	Linolenic		0.9	1.2	0.9	1.5
18:4w3	Octadecatraenoic		0.3	0.9	0.3	1.5
20:4w6	Arachidonic	1.3	0.2	0.2	0.2	0.2
20:5w3	Eicosapentaenoic		0.2	2.0		2.0
22:6w3	Docosahexaenoic		0.4	3.8	0.4	4.8

* Shultz and Ferguson (1974), ** Rouvinen (1991).

The degree of fatty acid saturation increases from the superficial tissues toward deeper fat reserves in both blue and silver foxes, illustrating the key insulation role of the subcutaneous adipose tissue in fox anatomy._

Bone

Smith (1941a) noted that dry fat-free ulna bones of Silver foxes (Vulpes fulva) contained 51.1+/- 2.7% ash.

Pup Growth

According to Ahlstrom et al. (2000), with blue foxes the average litter size is 8-12 and the body weight of the new born pups is about 65 grams. The pups increase their body weight 4-6 times during the first 3 weeks of lactation.

Milk Composition

TABLE 3.2 Fox Milk Composition— Silver Foxes, Blue Foxes

Nutrient	Colostrum(1)	Milk(2)	Milk (3)
Water	65.2	81.9	77.0
Fat	12.1	5.8	11.0
Protein	17.1	6.4	8.2
Lactose	4.0	4.8	3.0
Ash	1.5	1.0	1.0

(1) Anon. (1929), (2) Anon. (1930), Young and Grant (1931) and Spector (1956), (3) Rusanen and Valtonen (1991) and Ahlstrom (1992).

The experimental data of Rusanen and Valtonen (1991) with blue fox indicated that during early lactation the distribution of fatty acids was 27% saturated, 50% monounsaturated and 14% polyunsaturated. As the lactation period proceeded, the lipid composition became more saturated, evidence of a greater role of simple sugar resources as precursors for fatty acid synthesis.

Similar to other carnivores, the milk fat ofblue fox was characterized as principally C-16 and C-18 fatty acids. As would be expected, the composition of the milk reflected dietary fat composition such that the employment of 4% fish oil in the diet raised the level of C-20:1 and C-20:5 fatty acids significantly.

Studies by Ahlstrom and Wamberg (1997) and Ahlstrom et al. (2000) indicated that silver fox cubs had a milk intake of about 100 ml/cub/day at 17-21 days of age while blue fox cubs had a milk intake of 30-40 ml/cub/day at 13-15 days and 60-70 ml/cub/day at 19-21 days of age. Depending upon litter size, milk production of fox is as high as 500-700 ml/day. With blue foxes, milk production reached a maximum at 14-15 days with the energy output exceeding 770 kilocalories (3,200 kJ) day.

Digestive Tract

Anatomy

In polar foxes, the total digestive tract length is 210 cm with the small intestinal tract length of 177 cm (Szymeczko et al., 1992). Fox and mink have a similar intestine/body length, as noted in Table 3.3.

TABLE 3.3 Digestive Tracts of Animals

Animal	Intestine/Body Length Ratio
Horse	15/1*
Pig	14/1**
Rabbit	10/1**
Dog	6/1**
Cat	4/1**
Fox and Mink	4/1***

* Penelaik (1972). **Colin (1954) and Stevens (1977),*** Neseni (1935), Perel'dik et al. (1972), Elnif and Enggaard-Hansen (1988), Szymeczko and Skrede (1990).

With an intestinal/ body length less than dogs, it does not make common sense for fox ranchers to provide their animals dog food and expect top growth and fur development.

Food Passage Time

Although mink and fox have a similar intestinal/body length ratio of 4/1, the feed passage from mouth to anus is distinctly different. The mink employs an average of 3 hours with a wide range of 1 to 6 hours (Bernard et al., 1942; Wood, 1956; Howell, 1957; Neseni and Piatkokski, 1958; Sibbald et al., 1962; Anon., 1977; Bleavins and Aulerich, 1981; Charlet-Lery et al., 1981; Szymeczko and Skrede, 1990; Enggaard-Hansen et al., 1991). The average food passage time with the fox is 6-10 hours (Penelaik, 1972; Perel' dik et al., 1972; Anderson, 1988; Szymecko et al., 1992; Anon, 1996). Observations of Bernard et al. (1942) of 12 hours and those of Helgebostad (1976) of 24 hours are simply inconsistent with the later experimental studies.

Experimental studies by Szymecko et al. (1992) indicated that fox diets with eviscerated cod yielded 9.5 hours of feed passage while a ranch diet with the total protein resource as fish meal yielded only 7 hours of fox food passage. A study by Anon. (1996) indicated that the addition of wheat bran to fox diets decreased food passage time from 10 hours to 6.5 hours. Experimental work by Faulkner and Anderson (1991) indicated that the addition of 5% pectin to a meat-type diet for silver foxes yielded a faster rate of food passage.

The longer time of food passage in foxes relative to mink explains, in part, the higher digestibility capacity of foxes for carbohydrate and protein resources.

Intestinal Flora

Very little is known about the contribution of fox intestinal flora to the nutrition of the fox. Studies by Faulkner et al. (1992) indicated that oral doses of antibiotics increased the apparent digestibility of crude protein and ten individual amino acids. Antibiotic administration significantly reduced true digestibility of crude protein, but the reduction was only 1% and therefore may not be of any great significance in practical fox nutrition.

Physiology

Feedstuff Palatability

As early as 1948, Bassett et al. (1948) noted that in terms of experimental studies with a number of carbohydrate resources including bread, cooked potatoes, cookie crumbs and rolled oats, that the highest feed volume was with the foxes receiving the cooked potatoes. Studies by Chausow (1989) indicated that a powdered liver digest provided "on top" of fox pellets at a 0.1% level enhanced the palatability of the pellets for fox.

Digestive Physiology

Carbohydrate Resources

As early as 1927, an interest in the value of cooked cereal grains was expressed by G.E. Smith (1927). In general, both silver and blue foxes have a significantly higher capacity for the digestion of carbohydrates than mink. Thus studies by Bernard et al. (1942) indicated no difference in the capacity of mink and silver foxes to digest the carbohydrates present in cooked oats but the foxes had an added 10% digestive capacity in terms of the carbohydrates provided by rolled oats. Experimental work by Ahlstrom and Skrede (1995a) indicated that in ranch diets with low fat:carbohydrate ratios, i.e., relatively high levels of carbohydrate, the carbohydrates were more completely digested by foxes than by mink.

A summary of experimental data on the digestibility of carbohydrate resources is provided in Table 3.4.

TABLE 3.4 Digestibility of Carbohydrate Resources by Silver Foxes

Carbohydrates	Apparent Digestibility Rating
Barley, raw ground	72(1)
Oats, raw ground	73(1)
Wheat, raw ground	69(1)
Wheat Bran*	26(1)
Linseed meal	73(2)
Soya meal	84(2)
Bread Waste	88(3)

(1) Gunn (1948), (2) Smith (1942b), (3) Nordfeldt et al. (1955).

* The observation that silver foxes have a minimum capacity to digest wheat bran was confirmed by Bernard et al. (1942) and likewise in studies by Syzmeczko et al. (1996) with blue foxes.

The capacity of foxes to digest oatmeal may be impaired by overheating. Roasting oatmeal to 100 degrees Celsius somewhat depressed the digestibility of the proteins present and roasting to 200 degrees Celsius resulted in zero digestibility of the protein and fat of the meal and a slight reduction in the carbohydrates present, Nordfeldt et al. (1955). A study by Faulkner and Anderson (1991) with polar foxes indicated a very low digestibility of the carbohydrates provided by alpha cellulose, hemicellulose, oat bran, oat hulls, and pectin.

Commercial Carbohydratase Enzymes

Experimental work by Brzoowski et al. (2004) indicated that the employment of enzymatic preparations including alpha-amylase, beta-glucanase and xylanase on carbohydrate resources in blue (polar) foxes yielded only slightly better results, which were not significant statistically. A study by Valaja et al. (2004) indicated no significant differences in the digestibility of carbohydrates in barley between the untreated product and barley treated with lactic acid bacteria or beta-glucanase enzymes.

Fat Resources

Experimental data on the digestibility of animal and plant fats is presented in Table 3.5.

TABLE 3.5 Digestibility of Fat Resources by Blue and Silver Foxes

Fat Resources	Apparent Digestibility Rating	
	Silver Fox	**Blue Fox**
Frozen Animal Products		
Beef Hearts	98 (2)	
Beef Lips	99 (2)	
Beef Meat	98 (1)	
Beef Tripe (Rumen)	97 (2)	
Beef Manifolds	96 (6)	
Cow Udders	98 (2)	
Fish	96(6)	
Horsemeat	96 (1)	94(5)
Meatmeal	92 (6)	
Silkworm Pupa	98 (7)	
Whalemeat		91(5)
Rendered Fats		
Rapeseed Oil		96(3)
Capelin Oil		95(3,4)
Soybean Oil		90(4)
Beef Tallow		87(3)

(1)Inman and Smith (1941), (2) Inman (1941), (3) Rouvinen and Kiiskinen (1988),Rouvinen et al. (1988), (4) Skrede and Gulbrandsen (1985), (5) Titova (1959), (6) Gunn (1948), (7) Steger and Piatkowski (1959).

Studies by Rouvinen and Kiiskinen (1988) with blue fox indicated no synergistic effect of the combination of soybean oil and tallow on enhancing the digestibility of tallow. This same experimental work indicated that the digestibility of the total fat content of fox rations was dependent on the amount of stearic acid (C-l8:0) present in the specific fat resource with the following regressive equation:

$$\mathbf{y = 96.4 - 0.298x, \quad r = 0.963, R = 0.927 \text{ and } N = 39}$$

where y = apparent digestibility of the fat (%), x = stearic acid (%) in the fat of the diet, r = correlation coefficient, R = coefficient of determination and N = number of animals. The stearic acid content of fats does undermine the fat digestibility for the fox but not as markedly as it does with mink digestive physiology, (Jorgensen and Glem-Hansen, 1973). Studies by Ahlstrom and Skrede (1995) indicated that blue foxes had a higher capacity for fat digestion than mink and that with blue fox the fat digestibility rating did not decline with a decrease in fat:carbohydrate ratios as it did with the mink. An experimental report by Szymeczko et al. (1996) noted that the employment of wheat bran in polar fox diets to the point of yielding fiber concentrations above 1% led to a significant decline in fat assimilation.

Protein Resources—Protein Digestibility

Experimental data on the digestibility of animal and plant proteins is presented in Table 3.6.

TABLE 3.6 Digestibility of Protein Resources by Foxes

Protein Resourcs	Apparent Digestibility Rating	
	Silver Fox	**Blue Fox**
Animal Products		
Beef Hearts	95(1)	
Beef Lips	96(1)	
Beef Meat	97(2)	
Beef Tripe (Rumen)	95(1)	
Cow Udders	92(1)	
Fish	96(4)	
Horsemeat	93(2,3)	96(4)
Manifolds		93(4)
Sealmeat	93(6)	
Whalemeat		92(8)
Rendered Animal Products		
Bloodmeal	62*(4) -78(3)	
Casein	95(5)	
(Dried Cottage Cheese)		
Fishmeal	69*(4) - 88(3)	
Livermeal	88(3)	
Meatmeal	86(3)	
Silkworm pupa meal	95(7)	
Plant Products		
Linseed Oil Meal	81(3)	
Soybean Oil Meal	86(3)	

* Obviously an inferior product very likely related to excess dehydration programs.

(1) Inman, 1941; (2) Inman and Smith, 1941; (3) Smith, 1942a; (4) Gunn, 1948; (5) Titova, 1950; (6) Afanas'ev, 1957; (7) Steger & Piatkowski, 1959; (8) Titova ,1959.

A study by Szymeczko and Skrede (1991) on endogenous protein and amino acid excretion of polar fox is of interest. The experimental data indicated that the amount of endogenous protein was rather limited and thus appeared to have a minor effect on apparent digestibility of proteins by polar foxes fed normal diets. Their data on fecal amino acid composition after the introduction of "protein free" diets was comparable with data obtained earlier by Skrede et al. (1980).

A study by Ahlstrom and Skrede (1995) indicated that the apparent protein and amino acid digestibility ratings for blue fox were higher than that of mink. While fat:carbohydrate ratios did not have a significant effect on protein and amino acid digestibility ratings for the blue fox, for the mink decreasing fat:carbohydrate ratios resulted in declining digestibility ratings for both protein and amino acids.

Experimental data provided by Syzmeczko et al. (1996) indicated that with the employment of supplemental wheat bran in polar fox diets, as fiber levels rose above 1% dehydrated basis there was a significant decline in protein assimilation as well as a fall in digestibility ratings for all of the amino acids.

Protein Resources—Amino Acid Digestibility

Of major interest is the contrast in the protein and amino acid digestibility capacity of the fox relative to that of mink when a challenging protein resource such as meat-and-bone meal is provided as noted in Table 3.7.

TABLE 3.7 Protein and Amino Acid Digestibility Capacity of Mink and Fox Relative to Meat-and-Bone-Meal Digestibility Ratings

	Mink	Fox
	9 Months	15 Months
Protein	60	80
Essential Amino Acids		
Arginine	78	89
Histidine	53	88
Isoleucine	63	74
Leucine	66	75
Lysine	65	78
Methionine	64	71
Cystine	26	ND
Phenylalanine	73	81
Tyrosine	66	81
Threonine	57	78
Tryptophane	38	61
Valine	63	74
Non-Essential Amino Acids		
Alanine	67	81
Aspartic Acid	25	70
Glutamic Acid	55	74
Glycine	63	83
HO-Proline	70	87
Proline	67	83
Serine	64	78

Fox Manure

Annual fecal excrement per an adult female blue fox and silver fox is 36 and 22 kg/year (Aarstrand and Skrede, 1993). Note: Composition of fox fecal material is provided in the original publication.

4

Nutrient Requirements

Water

Without question, air and water are the most essential nutrients for animals. It has been estimated that a human being can live 4 minutes without air, 4 days without water and 40 days without food (assuming ample adipose (fat) tissue energy reserves). Water and salt are not ordinarily thought of as feedstuffs. They undergo no significant changes in the body but are used and excreted in the same form as ingested. However, in passing from infancy to maturity, substantial amounts of these nutrients are incorporated into the anatomy of the body.

Water is supportive of three major functions within the anatomy and physiology of the fox, i.e., (1) it gives structure and form to the animal's body, (2) it provides an optimum environment for the digestion of feedstuffs and for cellular metabolism, and (3) it is absolutely necessary for maintenance of optimum body temperature.

Water comprises about 70% of the total weight of an adult animal and is distributed with about 2/3^{rds} intracellular and 1/3rd extracellular compartments. The intracellular fluids exist within tissue cells while the extracellular fluids exist outside of the cells in two portions: an intravascular portion in the blood plasma and the extravascular portion outside of the blood circulation within the interstitial spaces. There is a basic difference in the composition of the intracellular and extracellular fluids. The intracellular fluids are high in potassium cations and low in sodium cations with significant levels of magnesium cations and phosphate and sulfate anions while the principal inorganic constituents of the extracellular fluids including the spinal fluid and the lymphatic fluids as well as the secretions of the digestive system and glands are high in sodium cations and low in potassium cations with bicarbonate anions. Note that blood has a "salty" taste. The water balance between the intracellular and extracellular compartments is maintained by a process termed osmosis (the movement of water is controlled by the concentrations of salts and colloidal particles including protein molecules). Blood albumin molecules play a key role in maintaining water balance between the blood circulation and the adjoining tissues. With liver damage and subsequent sub-optimum synthesis of blood albumins, the net result is an edema as the water moves from the blood to the interstitial spaces and accumulates in the body cavities. The inorganic cation and anion content of the intracellular and extracellular compartments are all isotonic with one another and therefore the loss of 1,000 mls of intestinal fluid by diarrhea causes the loss of an amount of sodium cations equal to that contained in an equal volume of blood.

Normally the marked difference in the inorganic ionic composition of the fluids present in the extracellular and intracellular compartments is well maintained because they are separated from one another by semipermeable membranes which although permitting the passage of water, glucose and metabolites, do not readily allow the passage of sodium and potassium cations.

Water Requirements

The three resources of water for the fox include (1) water content of the daily feed, (2) that provided by the water cup, and (3) metabolic water which results from animal physiology as the direct result of metabolic processes. These include (a) the oxidation of simple sugars from carbohydrates, fatty acids and glycerol from fats, and amino acids from protein resources to yield carbon dioxide and water and ammonia in the case of amino acid catabolism; and (b) dehydration processes involving polymerization of sugars to yield more complex carbohydrates, the polymeric program involved in polypeptide and protein synthesis, as well as the biochemistry involved in creating triglycerides (simple fats) from glycerol and free fatty acids. In terms of the mink, Farrell and Wood (1968) have estimated that water requirements of the female mink are being met by:

Feed Resources	**66%**
Fluid Resources	**14%**
Metabolic Water	**20%**

Experimental studies by Dille et al. (1998) indicated that there is a wide individual variation in water consumption of farmed foxes, with a daily water consumption of 100 - 600 mls./day. Studies by Moe et al. (2000) indicated that plasma osmolality and urea concentration increased continuously throughout a 4 day period of water deprivation, even when the foxes were fed a standard ranch feed with a high water content of 60-70%. Such an increase in plasma osmolality indicated that the fox's ability to concentrate urine in order to compensate the water limitation was exceeded.

Energy

Energy is not a nutrient but a property possessed by fats, carbohydrates, and proteins. Whereas the major function of proteins is to supply amino acids for the construction of enzymes, muscles, organs, blood, bone, and fur, etc., the primary function of carbohydrates and fats is as energy resources with a secondary role in the composition of cell structures.

A full appreciation of the key role of energy in fox nutrition and nutritional physiology requires a clear understanding of multiple definitions directly related to the biology and chemistry of the fox:

1. **Energy in Physical Science**—the capacity to do work;
2. **Energy in Chemical Science**—caloric value of a feed product as determined by combustion within a bomb calorimeter, i.e., the Gross Energy value of a feedstuff;
3. **Energy in Biochemical Science**—caloric value of a feed product as actually provided to a living animal involving the processes of digestion through nutritional physiology and thus capable of being sub-divided into Energy—Digestible = Gross Energy - fecal energy loss;
4. **Energy considered Metabolically = Digestible Energy** — urinary energy and gaseous energy loss, i.e., M.E. = Gross Energy—(fecal energy + urinary energy + gaseous energy);
5. **Energy considered as Net = Metabolic Energy**—heat energy loss. The Net Energy represents the fraction of the gross energy that is actually utilized by an animal for productive purposes. While the Net Energy is the most precise estimate of a feed's energy value, it is not of practical employment in modern fox nutrition and physiology because of the difficulty of measuring heat losses. Thus Metabolic Energy (ME) is employed in calculations of a fox's feed intake and the requirement of nutrients;
6. **Energy Balance**—Negative = wherein an animal metabolizes its energy reserves, primarily fat, to maintain the processes of life. This energy is illustrated with fox on a restricted feed intake prior to breeding for proper conditioning for breeding or during the lactation period when the extra energy output as energy rich milk may exceed the nursing vixen's capacity for adequate energy intake;
7. **Energy Balance**—Positive = wherein an animal's energy intake exceeds its daily needs and the excess energy is deposited primarily in the form of fat (adipose tissue) and protein (body structural components including fur;
8. **calorie**—the amount of heat energy required to raise the temperature of one gram of water by one degree Celsius measured from 14.5 to 15.5 degrees Celsius–the room temperature of the European laboratory in which the original work was conducted;

9. **Calorie**—the kilocalorie = 1,000 calories, i.e., Calorie with a capital C represents an energy value 1,000 times the energy value of a calorie with a lower case letter;
10. **Joules**—one calorie = 4.184 joules. Joules are the energy unit to be employed in the international scientific community in the 21st century;
11. **Metabolism**—animal = total of all biochemical processes taking place within an animal's physiology;
12. **Metabolism—Basal** = the lowest rate of metabolism of an animal in a state of complete rest (in the case of the fox when it is asleep), and in a thermal neutral environment and in a post-absorptive state, i.e., the amount of energy required for the involuntary work of the body including the function of the various organs including the heart, kidneys, and lungs and the maintenance of body temperature (via oxidative reactions in resting tissue, especially in the maintenance of muscle tone) and a physiological environment without involving feed digestion processing. Thus the Basal Metabolism is the result of the energy exchanges taking place in the cells of the body, i.e., without the involvement of body fat, extracellular fluids, and the bone marrow. It is of interest to note that the sodium pump activity involving sodium- potassium ATPase in the muscles and liver represents as much as 10-20% of the total basal energy expenditures;
13. **Basal Metabolism Rate (BMR)**—Energy requirements of an animal in a physiological environment of minimal energy expenditure;
14. **Maintenance Metabolic Rate (MMR)**—Energy expenditure under conditions of no weight gain or loss under normal conditions of activity and environmental temperature. MMR must be met before any production functions can be accomplished including pregnancy, lactation, kit growth, and fur development;
15. **Metabolic Weight = Metabolic Body Size (MBS)**—$BW^{0.75}$ (body weight kilograms).

All factors considered, an animal's highest priority for nutrient intake is for energy; all other needs such as a requirement for amino acids, simple sugars, and essential fatty acids for cell synthesis become secondary. Thus, for example, if a fox's diet provides an energy level from carbohydrates and fats which is insufficient for its daily energy needs, the animal will simply convert amino acids into energy as a higher need than the synthesis of body structures including fur.

Fox—Nutrient Resources–Metabolic Energy Values

Until specific experimental data is available on the Metabolic Energy (ME) values of nutrients for modern fox nutrition, the ME values of protein, 4.5; fat, 9.5; and carbohydrates, 4.0 are recommended, as noted in the National Research Council (U.S.) Bulletin Number 7, Nutrient Requirements of Mink and Fox, Travis et al. (1982).

Maintenance Metabolic Rate—MMR–MEm

The MEm of an animal includes all the physiological processes by which the species maintains its body without any change in animal weight or body composition. These can be divided into special energy requirements for:

1. **Basal Metabolism;**
2. **Thermoregulation;**
3. **Activity Pattern–Locomotion;**
4. **Heat of Nutrient Metabolism.**

Brief commentary is warranted on point 4. Heat of Nutrient Metabolism. The assessment of the BMR of an animal is achieved with the animal at complete rest and in a post-operative state, i.e., a physiological environment without the energy enhancement related to feed digestive processing. The rise in heat output following a meal represents energy released during the digestion, absorption, and assimilation of the nutrients in feedstuffs. This is termed the heat increment (HI) of feeding or heat of nutrient metabolism. The amount of heat energy released, HI, represents less than 20% of ME available from fats but as high as 50% of ME available from proteins. The HI directly related to protein intake used to be termed Specific Dynamic Action, SDA.

Fox Metabolic Energy for Maintenance, MEm

The maintenance energy requirements of an animal must be met before any productive functions can be accomplished. For many animals, the MEm is about twice the BMR (Brody, 1945; Kleiber, 1961).

TABLE 4.1 Metabolic Energy Requirements for Maintenance, MEm for Silver Fox*

Year	MEm Cal/kg/day	Researchers
1927	111-121	Palmer (1927)
1931-1934	95-100**	Smith (1935)
1942	112	Hodson and Smith (1942b)
1958	121	Mamaeva (1958)

* Calculations based on ME values/gram of carbohydrates, 4.1; protein, 4.1; and fat, 9.3.

** Size very likely smaller in terms of the early years of commercial fox ranching.

The Hodson and Smith experimental data on the MEm, Cal/kg/day for mink is 209 and for fox is 112. At first glance, it appears that on the basis of body weight, the MEm requirement of the mink is almost twice that of foxes, by logic requiring a relatively higher feed volume/day. Obviously, this does not make common sense to any fur farmer involved in the production of both fox and mink pelts.

The relatively low MEm value for fox compared to mink is meaningless in terms of considering energy requirements on a unit weight/day basis inasmuch as the energy needs of animals in terms of basal metabolism and for maintenance are more nearly proportional to surface area such that the greater the surface area of an animal, the greater potential for body energy loss as heat. In terms of MEm relative to square meters of body surface area, the Hodson and Smith data for the mink is 1830 and for the fox 1948. Obviously, mink and fox are marketed on the basis of pelt size and not on the basis of body weights.

The surface area of an animal can be estimated by the formula of Lusk, $S = Kw^{2/3}$ wherein S = surface area, w = live weight and K is a constant for each species. The value for the dog is 0.103, which can be applied to the fox involving only a minor error. Small, average, and large animals of the same species have approximately the same energy requirements in terms of animal surface area, but in terms of body weight, the smaller animals will have a higher energy requirement while the larger animals will have a lower energy requirement than average.

MEp—Metabolic Energy–Production

The production functions of the fox include reproduction/lactation, growth, and fur production. Each phase of the ranch year requires special consideration relative to any survey of the energy requirements of the fox, MEp, including:

1. Pre-Breeding;

2. Reproduction;

3. Lactation;

4. Growth; and

5. Fur Development.

MEp—Pre-Breeding Phase

At the present time, to my knowledge, no experimental data is available on the MEp for the pre-breeding phase.

MEp—Reproduction Phase

A study by Ahlstrom and Skrede (1997) is of interest. Their research work indicated that blue fox females with high initial weights and high energy supply—2,250 kJ/day (538 kcal/kg/day)—in the early phase of the reproduction period resulted in several cases of appetite loss and very low feed intake during gestation, resulting in kits with low birth weights and poor viability.

TABLE 4.2 Energy Requirements Silver Foxes—Pregnancy

Phase	Heat Production Kcal/kg/24 hours - Range	
Early Pregnancy—1st week	55-60*	78**
Mid Pregnancy—3 weeks	58-60*	
Late Pregnancy—5 weeks	64*	91**

*4 fetuses, **7 fetuses Perel'dik et al. (1972)

MEp–Lactation Phase

Studies by Abramov and Poveckij (1955) with blue foxes resulted in a recommendation of a 650 kcal/kg/day for the lactating mother supplemented according to the number and age of the young.

TABLE 4.3 Energy Supplementation for Lactation—Blue Foxes

Litter Age	0-10	10-20	20-30	30-40	40-50	50-60
Daily Allowance /Pup	50	115	200	300	350	400

MEp–Growth Phase

Experimental data on the energy requirements of the fox during the growth period are provided in the following tables.

TABLE 4.4 Silver Fox Kit Weights and Energy Requirements

Age/Weeks	Kit Weight/Grams	Metabolic Energy Requirements* kcal M.E.
0	100	—
1	225	11
2	365	74
3	515	120
4	680	166
5	880	225
6	1,130	312
7	1,400	378
8	1,650	428

* Rimeslatten (1978)

TABLE 4.5 Male Kit Weights and Energy Requirements

	Age in Months	Silver Fox			Blue Fox	
Initial Weight kg		kcal ME			Initial Weight kg	kcal ME
		*	**	***		**
1.8	1,4	280				
2.3	1.8		390		2.3	560
2.8	2.5	450				
3				470		
3.4	3.0		510		3.8	710
3.8	3.6	540				
4				570		
4.5	4.1		550		5.0	690
4.8	4.4	620				
5				680		
5.6	5.0		570		5.8	630
5.8	5.1	640				
6				600		
6.7	5.8		490		6.1	590
6.8	5.8	580				
7				550		
7.8	6.0	430			6.2	520
7.8	6.2	500				
8				500		
8.8	6.5	490				

* Rimeslatten via Travis et al. (1982), **Penelaik (1975), ***Mamaeva (1958)

TABLE 4.6 Silver Fox Female Kit Weight and Energy Requirements*

Age/Weeks	Initial Weight	kcal ME
7	1.4	260
11	2.3	410
15	3.2	490
19	4.0	550
23	4.6	580
27	5.1	510
31	5.4	430
35	5.5	410

* Rimeslatten via Travis et al. (1982).

TABLE 4.7 Heat Production of Fasting Young Foxes

	Silver Foxes		Blue Foxes	
Month	Age -Months	Cal/kg.day	Weight	Cal/kg.day
July	2	105	1680	123
August	3	82	2660	104
September	4	63	3260	91
October	5	59	3680	81
November	6	56	4010	75
December	7	50	4560	64

Perel' dik et al. (1972).

MEp–Year Summary

TABLE 4.8 Energy Requirements—Fox Ranch Year

Month	Adult Females				Adult Males	
	Silver		Blue		Blue	
	Kg	Kcal/kg	Kg	Kcal/kg	Kg	Kcal/kg
January	6.4	68	6.4	70	8.0	62
February	5.7	75	6.2	66	7.6	59
March	5.5	105	5.8	75	7.2	64
April	5.6	104	5.4	97	6.6	92
July	4.8	113	4.8	119	5.8	116
August	5.1	115	5.0	122	6.1	118
September	5.6	107	5.4	125	6.6	122
October	6.0	92	5.9	108	7.3	107
November	6.3	84	6.3	97	7.8	93
December	6.5	72	6.5	78	8.0	75

Perel'dik (1972) as modified in Travis et al. (1982).

Protein and Amino Acids

Johannes Mulder

Proteins are polymers consisting of as many as twenty two different units termed amino acids. Proteins are the main components of the muscles, organs, and endocrine glands. They are the major constituent of the matrix of bones, teeth, skin, nails, and hair as well as the blood components including hemoglobin and the plasma polymers responsible for the regulation of osmotic pressure and in the maintenance of water balance in the body. The blood antibodies as well as enzymes and many hormones are also protein in nature. All the proteins present in the anatomy and physiology of an animal are in a constant state of flux, i.e., a turnover related to degradation and synthesis, as is well illustrated by human blood cells with a life span of about 120 days. Even though some of the constituent amino acids are resused, the recycling metabolic processes are not completely efficient. Also, some of the amino acids are used for energy and some proteins are lost from the body. In the case of growing, furring, or pregnant animals, additional body tissue is being synthesized. Since animals cannot synthesize all of the amino acids they require, as plants can do, they require an outside resource of protein (amino acids) to survive and reproduce. In terms of animal nutrition, proteins also serve as an energy source for fur animals, providing a metabolic energy (ME) value of 4.5 kilocalories/gram.

In 1838, a Dutch chemist, Johannes Mulder, described certain organic material as

> *"unquestionably the most important of all known substances in the organic kingdom. Without it no life appears possible on our planet, through its means the chief phenomena of life are produced."*

Berzelius, a contemporary of Mulder, suggested that this complex nitrogen-bearing substance be called "protein" from the Greek word *proteios* meaning "take the first place or primary."

Inasmuch as the fox is primarily a carnivore, it is obvious that protein resources are the primary topic of consideration in the planning of commercial fox ranch programs. For modern fox nutrition management, without question, protein is the key nutrient since (1) pelt production requires more protein than any other nutrient and (2) in feed economics, protein is the most expensive nutrient resource required per unit and in terms of total units per pelt marketed.

Animals do not require protein of itself, but actually require the individual amino acids present in the feedstuff protein, all of which are essential units for the synthesis of protein polymers required in the anatomy and physiology of animals. However, animal bodies can synthesize more than half of the required amino acids through metabolic products of glucose and "extra" nitrogen resources, including amino acids provided in the body's physiology which are in excess of actual requirements of the animal at that specific time period. This process is termed transamination. Thus dietary proteins are considered in terms of their biological value as directly related to their capacity to provide an optimum measure of both essential and non-essential amino acids. The essential amino acids are considered to be indispensable inasmuch as the animal body can not synthesize them, while the non-essential amino acids (actually required by an animal's physiology) are termed dispensable inasmuch as the animal physiology is capable of providing them via metabolic processes.

The amino acid requirements of the fox can be classified as follows:

Essential: Arginine, Histidine, Isoleucine, Leucine, Lysine, Methionine, Phenylalanine, Taurine, Threonine, Tryptophan, and Valine.

Semi-Essential:* Cystine and Tyrosine.

Non-Essential: Alanine, Aspartate, Glutamate, Glycine, Hydroxy-Proline, Norleucine, Proline, and Serine.

*Amino acids without an absolute dietary requirement but in the presence of which the animal's diet can minimize the requirement of a corresponding essential amino acid. This is accomplished by modification of these semi-essential amino acids by metabolic processes of the animal, thereby providing a "fill-up" capacity. Such action is illustrated by the conversion of tyrosine (non-essential) to phenylalanine (essential). The key role of cystine (non-essential) in fur synthesis is minimizing the dietary requirement for methionine (essential) for metabolic conversion to cystine in the critical fur development phase of the fox ranch year.

Proteins are very large molecules obtained via the combination of multiple amino acids (monomers) in a linkage termed a peptide bond (secondary amide) to yield structures with hundreds of amino acid units. Smaller chains of amino acids are termed polypeptides, well illustrated by hormones such as human insulin with 66 amino acids and oxytocin with nine amino acids in a cyclic structure.

Physiology

For the most part, protein molecules and large polypeptides cannot be absorbed by the intestinal mucosa in any appreciable degree and hence must be broken down (hydrolyzed) by enzymes in the stomach and small intestine to yield free amino acids and small peptides. The exception is colostrums, the immature milk provided at birth which is unique in terms of high protein levels related in part to immunoglobulins. These immunoglobulins (congenital antibodies) are absorbed intact and provide the newborn with transient protection in the form of passive immunity, for example to distemper, for up to 8-10 weeks.

Protein digestion is initiated in the stomach where hydrochloric acid not only denatures the proteins so that they are more susceptible to enzymatic action but also activates the inactive pepsinogen structure (secreted into the stomach by the direct physiological reaction to food intake) to the active form of pepsin. Pepsin brings about a breakage (hydrolysis) of specific peptide linkages to begin the process of digestion of the protein. In the small intestine, pancreatic secretions include two inactive proteolytic enzymes, trypsinogen and chymotrypsinogen. Trypsinogen is activated to trypsin by enterokinase, an enzyme from the intestinal wall. Chymotrypsinogen is activated by trypsin to chymotrypsin. These active enzymes continue the process of hydrolytic cleavage of the protein by attacking specific peptide linkages. The final hydrolysis of proteins and polypeptides is completed by a group of enzymes termed peptidases which are secreted from the intestinal mucosa. The end products of digestion are amino acids, dipeptides and tripetides, which are absorbed by the gut mucosa. Hydrolysis in the brush border and also in the cystosol of the mucosal cells insures that most of these small peptides will yield free amino acids for the portal circulation leading to the liver. The liver maintains control over the flux of amino acids via a number of reactions including catabolic reactions to yield energy and urea as a by-product, synthesis of specific blood proteins, and the release of free amino acids into the general blood circulation.

Protein Resources—Biological Value–Quality

An animal's requirement for protein for each phase of the fox ranch year will be directly dependent on the biological value of the protein resources provided the animal. With protein feedstuffs of high biological value or quality, a minimal level of protein will be required. With protein feedstuffs of lower quality greater and greater quantities of protein will be required to meet the animal's amino acid needs for top performance.

Two factors determine the biological value or quality of a given protein feedstuff for an animal, i.e., (1) amino acid pattern and (2) availability of those amino acids to the digestive processes of the animal. The world's standard for highest biological value is whole egg protein with (1) an amino acid pattern similar to animal requirements and (2) high digestibility by all animals. On the other hand, chicken feet are illustrative of a low quality protein resource inasmuch as (1) it possesses an amino acid pattern which is inconsistent with the actual amino acid requirements of most animals as a consequence of relatively low levels of certain amino acids including tryptophan and (2) has a very low digestibility for animals, as low as 52% for mink (Leoschke, 1959).

Carefully processed fish meals have a good amino acid pattern and high digestibility for foxes. However, overheated fish meals, especially those with a moisture content below 6%, provide a sub-optimum amino acid pattern to the fox inasmuch as excessive heating destroys lysine and bonds arginine in a structure unavailable to the digestive processes of animals (Allison, 1949). Tryptophan and the sulfur amino acids, cystine and methionine, are especially sensitive to destruction during the dehydration of protein feedstuffs (Varnish and Carpenter, 1975).

Experimental data on the protein requirements of the fox for different phases of the ranch year show the need for diets with a good proportion of protein feedstuffs of high biological value for the fox. With an even higher percentage of high quality protein resources, lower levels of protein quantity may be effective in providing top performance of the fox. At the same time, as higher levels of protein feedstuffs of lower biological value for the fox are used, higher levels of protein will be required to meet the amino acid requirements of the fox for that specific period of the ranch year.

Protein Requirements–Quantity

The earliest statement on the protein requirements of foxes is in a Canadian report by Smith (1935). The recommended protein requirement for silver foxes was 7-8 grams of protein per kilogram of body weight. This experimental work was followed by Harris et al. (1951) with silver fox: their research data supported a recommendation of a minimum of 25% protein as dry matter (DM) basis (ca. 23% ME) for the growth period and 19% protein on a DM (ca. 17% ME) for the furring phase. This data is certainly in question in terms of our current understanding of the protein requirements of fur animals, as multiple studies indicate a higher protein requirement for top fur development relative to the protein needs for a fur animal for top weight gains during the late growth phase (July-August).

Relative to a comparison of the protein requirements of silver and blue fox on current practical ranch diets, the observations of a Swedish fur rancher, Raymond (1980) with mink (25,000) and fox (1,000) are of interest. He noted that silver foxes require more meat than blue foxes, silver foxes requiring a ranch diet with 75% meat and blue foxes needing a dietary regimen with only 40% meat.

In the paragraphs to follow, note that the protein requirements of foxes are significantly lower than those of mink for the same period of the ranch year. This finding is directly related to the basic fact that foxes (1) have a more efficient digestive system and (2) use protein primarily for tissue synthesis and not as a major energy resource, as in the case of the mink.

All factors considered, methionine is the most limiting amino acid in fur animal diets, including that of the mink (Dahlman et al., 2004). Thus in the experimental studies acknowledged in the paragraphs to follow, the requirement for methionine supplementation of fox diets during both the late growth and fur development phases of the fox ranch year occurs in diets designed for minimum protein levels for top performance.

Early Growth—7–16 Weeks Harris et al. (1951a) found that more than 40.7% protein (dehydrated basis) was required to attain maximum nitrogen storage in silver fox pups between 7 and 23 weeks of age. However, growth of foxes on a diet containing only 24.5% protein was equivalent to that of foxes on higher protein levels. Rimeslatten (1976b) has observed that blue fox pups raised on diets containing less than 28-30% ME from digestible protein attained normal body weight but reduced body length.

Late Growth—16 Weeks to Pelting Experimental studies by Rimeslatten (1976a, 1976b) indicated that 25% of ME as protein was ample for blue fox from 16 weeks to pelting. More recent experimental work by Dahlman et al., (2003) indicates that protein levels as low as 15% may be satisfactory for top growth performance of the fox during July-August, provided that the diets are supplemented with methionine.

Fur Development Rimeslatten (1976a, 1976b) recommended a level of 25% ME for the period from 16 weeks to pelting for blue foxes. Fur development quality was not significantly enhanced with protein levels in the range of 26-28% ME. More recent experimental work, especially research diets supplemented with methionine, indicate that 20-22% of ME as protein is quite satisfactory for the fur development of blue foxes, a recommendation, obviously based on fox diets with significant levels of protein resources with high biological value for the fox. (Berg et al., 1982; Typopponen et al., 1987; Hansen et al.,1990; Dahlman and Valaja , 2003; Dahlman et al., 2003; Dahlman, 2005; Koskinen et al., 2005a , 2005b).

Reproduction Rimeslatten (1976b) noted that estrus, breeding, and reproduction of blue foxes were not significantly influenced by alterations in protein concentrations ranging from 25 to 40% ME. A slight reduction in litter size at birth was observed when vixens were fed protein levels below 31-32% ME; however, this difference was not considered to be significant. Hence, Rimeslatten's recommendation of a minimum protein level of 30% ME for the reproduction period.

A more recent study by Polonen and Dahlman (1988) with blue fox appears to indicate that a higher protein, lower carbohyrate nutritional regimen may be more favorable for enhancing reproduction.

TABLE 4.9 Nutrient Ratios and Reproductive Performance of Blue Fox

Nutrient Ratios—% ME	Kits/Mated Female	Infertility—%
Protein, 33; Fat, 41; and Carbohydrate, 26	5.6	35
Protein, 40; Fat, 38; and Carbohydrate, 22	6.5	29

Obviously, the biological value of the protein resources in the fox diet has a major role in determining the reproductive performance of the animal. It is apparent from the table that with the protein resources of the control diet, extra protein was required for top reproductive performance.

Lactation

Rimeslatten's experimental work (1976b) indicated that reduced weight gains of fox pups were noted when their mothers received diets containing protein levels below 30% of ME.

Summary—Protein Requirements of Fox–Ranch Year

The Enggaard-Hansen et al. (1991) report of the Nordic Association of Agricultural Scientists' recommendations for fox nutrition in terms of the distribution of Metabolic Energy is Table 4.10.

TABLE 4.10 Recommended Distribution of Metabolic Energy (%) From Protein, Fat, and Carbohydrates in Fox Diets

Period	Protein	Fat	Carbohydrates
December-Whelping	35	20-45	35
Whelping-7/15	37	35-50	25
7/15-August	28	35-55	30
September-Pelting	26	35-55	35

For silver foxes, the National Research Council (USA) (Travis et al., 1982) reports in Table 4.11.

TABLE 4.11 Nutrient Recommendations of Silver Foxes Amount per 100 kcal Metabolizable Energy

Period	% Metabolized Energy
Growth—7–23 Weeks	28-30
Growth—23 Weeks to Maturity	25
Maintenance	22
Gestation	30
Lactation	35

Amino Acid Requirements

Very little experimental work has been done on the amino acid requirements of the fox. Recent studies by Dahlman et al. (2004) and Dahlman (2005) on the amino acid requirements of blue fox is of major interest. Employing nitrogen responses of the fox as criteria, they determined that the optimum amino acid pattern for recently weaned foxes is as provided in Table 4.12.

TABLE 4.12 Optimum Amino Acid Pattern for Blue Foxes as Expressed Relative to Lysine-100

Lysine	100
Methionine+Cystine	77
Threonine	64
Histidine	55
Tryptophan	22

Without question, in terms of fur development, methionine is likely to be the most limiting amino acid as it is a major metabolic resource for cystine, the key amino acid component of fur. Thus the ratio of methionine + cystine would likely be even higher than that provided with this experimental data, achieved with recently weaned foxes in the early growth period post lactation.

Methionine

Studies by Dahlman and Valaja (2003) indicated that fox are able to metabolize both D- and L-methionine on an equal basis in contrast to mink which show a very limited capacity to utilize D-methionine (Elif and Hansen, 2005).

Lysine

Multiple studies including Dahlman et al. (2002a, 2002b), Dahlman and Valaja (2003) and Dahlman (2005) indicate that L-lysine supplementation related to experimental work on the protein requirements of the fox provided decreased early growth of fox pups and impaired fur density. Very likely the adverse effects of lysine supplementation was directly related to the basic fact that lysine and arginine have a common absorption site on the intestinal wall. Thus extra lysine levels in the fox diet may result in reduced absorption of arginine, a key amino acid for fox nutrition and especially for fur development.

Taurine

There are significant field observations that indicate that fox, like the cat, have a specific requirement for taurine. In the cat, the requirement for taurine is directly related to the fact that cat metabolism pathways do not include a mechanism for converting cystine sulfinic acid to taurine (Hayers et al., 1975). Observations of 1987 indicate significant losses of fox pups on a specific commercial pellet program employing high levels of meat meal and relatively low levels of fish meal. Common pathology of all the fox pups was cardiac sympathy, considered to be directly related to a taurine deficiency (Kittleson, 1987; Barkoll, 1987). It is important to note that fish meals provide significantly higher levels of taurine compared to meat-and-bone meal.

Tryptophan

Experimental studies by Rouvinen et al. (1999) indicate that tryptophan supplementation of fox diets can have a positive effect on the behavior response of the animals. Higher levels of tryptophan have a potential for psycho-pharmacological properties inasmuch as tryptophan is an amino acid precursor of the neurotransmitter serotonin, which has been demonstrated to reduce fear and enhance exploratory behavior in the silver fox female. This is attributed to the fact that the female fox is more sensitive to the imbalance between tryptophan and other large neutral amino acids. Tryptophan supplementation of fox diets results in an increased serotonin synthesis. The brain serotoninergic system plays a significant role in the process of domestication of the silver fox. Tryptophan supplementation of ranched fox dietary programs may provide a tool for management of undesired behavior in breeding fox females.

Fats and Essential Fatty Acids

Fats are fatty acid tri-esters of glycerol (triglycerides) which belong to a class of biochemicals known as lipids which also includes monoglycerides, diglycerides, phospholipids (cephalin and lecithin), sphingolipids, glycolipids, cholesterol, and phytosterols. The sterols can exist as free alcohol structures or esterified with long-chain fatty acids. The Triglycerides are usually called fats if solid and oils if liquid at room temperature.

Inasmuch as the fox is a carnivore, fats play a more significant role in meeting its energy requirements than they do in omnivores such as the pig or herbivores such as the rabbit. Inasmuch as no specific fox energy values for the energy nutrients have been reported in the scientific literature, the data provided in Table 4.13 may be used in the calculation of the energy content of fox diets.

TABLE 4.13 Nutrient Metabolizable Energy Values for Mink

Nutrient	Metabolizable Energy Value/Kilocalories/Gram
Fat	9.5
Protein	4.5
Carbohydrates	4.2

Special attention must be given to the basic fact that the fatty acids present in triglycerides can be divided into two classifications, i.e., saturated fatty acids with only single bonds between the carbon atoms and unsaturated fatty acids with both single and double bonds between the carbon atoms, as illustrated in Table 4.14.

TABLE 4.14 Common Fatty Acids in Fox Nutrition

Fatty Acid	Designation*	Major Resources
Myristic	14:0	Animals and Plants
Palmitic	16:0	Palm Oil, Animals, and Plants
Palmitoleic	16:1	Animals and Plants
Stearic	18:0	Animals, especially cattle
Oleic	18:1	Olive Oil, Animals, and Plants
Linoleic	18:2	Animals and Plants
Linolenic	18:3	Linseed Oil and Horse Fat
Cetoleic	22:1	Fish
Arachidonic	20:4	Animals and Fish
Eicopentaenoic	20:5	Fish
Docosahexanoic	22:6	Fish

* 16:1 refers to a fatty acid with 16 carbon atoms and one double (alkene) bond.

Fatty acids without alkene bonds have no room for additional hydrogen atoms and are therefore termed "saturated" fatty acids while fatty acids with one or more alkene bonds have the capacity to add hydrogen or other atoms to their structures and are therefore termed "unsaturated." Fats with relatively low melting points, termed oils, generally have a higher content of short-chain fatty acids as found in butter and palm oil or contain a significant level of polyunsaturated fatty acids as found in vegetable oils. Softer animal fats such as poultry fat may contain a relatively high level of oleic (18:1) and linoleic (18:2) acids, as Table 4.15 illustrates.

TABLE 4.15 Composition of Animal, Plant and Fish Fats

	Pork	Chicken	Horse	Corn	Soya	Linseed
References	(1)	(1)	(1)	(2)	(3)	(3)
14:0	1.7	1.0	5.8			
16:0	28.6	23.0	30.2	10.6	10.7	5.0
16:1	2.6	5.6	6.6	0.1	0.2	
18:0	17.8	7.3	5.6	1.8	3.7	4.9
18:1	35.7	40.1	27.0	27.3	24.8	20.2
18:2	8.4	20.9	10.4	53.2	53.7	16.0
18.3	0.7	1.1	10.2	1.2	7.2	52.5
20:1	1.9	1.0	0.5	0.1		
20:4		0.3	1.4			

Fat	Capelin	Herringmeal	Salmonmeal
References	(3)	(4)	(4)
Fatty Acid			
14:0	7.1	6.0	5.5
16:0	11.9	14.4	16.4
16:1	10.1	5.7	7.6
18:0	1.3	2.2	3.7
18:1	13.4	14.7	17.9
18:2	1.4	1.8	3.3
18:3	0.7	0.7	0.9
18:4	3.9	1.2	1.7
20:1	12.8	12.7	7.2
20:4	0.5	0.9	2.0
20:5	9.9	6.0	7.5
22:1	3.4		
22:5	0.8	0.9	2.6
22:6	9.4	8.90	9.9

(1) Yu and Sinn Huber (1967), (2) USDA (2007), (4) Kaekela et al. (2001), (4) Fleming (1999).

In a chemical process termed catalytic hydrogenation, chemists can convert vegetable oils like corn oil or soybean oil containing high levels of linoleic acid (18:2) to stearic acid (18:0) and thereby yield solid fats similar to margarine. This same process of hydrogenation can also occur biochemically in the rumen of beef cattle and dairy cows, causing the highly saturated nature of beef fat and cow's milk. An exception to this observation that most animal fats are highly saturated is horse fat with a high content of linolenic acid (18:3).

In addition to being a significant energy resource for the fox, fats act as physiological carriers of feed flavors and the fat-soluble vitamins, A, D, E, and K. This process includes absorption from the intestines into the lymphatic system and final passage to the liver and the total blood circulation. Fats are also the only source of the essential fatty acids required for the synthesis of prostaglandins, important for the regulation of body metabolism. They provide a major energy storage nutrient as adipose tissue, important in the winter months when feed resources may be limited. Fats also provide thermal insulation and physical protection for the internal organs. Like carbohydrates, they have a "sparing" effect on the protein requirements of animals, that is, optimum levels of both fats and carbohydrates in the fox diet can lead to minimal protein levels required for top performance of the fox in all phases of the fox ranch year.

Physiology

Digestion and absorption of fat into the fox's lymphatic pathway to the thoracic duct and finally to the blood circulation is facilitated by the combined actions of the enzyme pancreatic lipase, bile salts and phospholipids, such as lecithin, and the peristaltic action of the small intestine. The action of pancreatic lipase is to bring about hydrolysis of the ester linkages of triglycerides to yield free fatty acids (FFA), diglycerides, and 2-monoacylglycerides. The endoplasmic reticulum of the intestinal wall is the site of resynthesis of triglycerides from the absorbed products of digestion. The combination of lipoprotein apopeptides with these triglycerides yields chylomicrons which are then sent into the blood circulation via the lymphatic pathway and the thoracic duct. Once within the blood circulation, the lipoprotein lyase releases free fatty acids (FFA) for uptake by the adipose (fat) tissue, the liver, and other organs.

Fatty Acid Metabolism—Omega-3 Structures

Multiple experimental studies indicate that relative to the mink, both blue fox and silver fox fatty acid metabolism is impaired relative to the oxidation of long chain omega-3 polyunsaturated fatty acids including eicopentaenoic acid (EPA), C20:5 omega-3; Cetoleic acid, C22:1 omega-3; and docosahexaenoic acid (DHA) C-22:6 omega-3 (Shultz and Ferguson, 1974; Rouvinen, 1987; Rouvinen and Kisskinen, 1989; Rouvinen et al., 1989; Rouvinen, 1991, 1992; Rouvinen et al., 1992; Ahlstrom and Skrede, 1995d).

In silver fox livers, the amount of DHA was even higher than in those of blue foxes. In addition, it was noted that cetoleic acid was more prominent in silver fox heart muscle and subcutaneous fat than in corresponding tissues of the blue fox. Even with equal quantities of EPA and DHA in the dietary regimen of the fox, the accumulation of DHA was, in nearly in all cases, much more dramatic than that of EPA. The 22 carbon chain of DHA was simply more difficult to metabolize than the shorter chain EPA with only 20 carbons.

Higher concentrations of these omega-3 PUFA in fox body fat depots may cause stress on the anti-oxidant capacity of the tissues. In severe cases of lipid peroxidation and sub-optimum dietary levels of vitamin E, the net result can be anemia, poor growth, depigmentation of the fur, yellow fat, and muscle dystrophy (Helgebostad, 1976). In their natural environment, blue and silver fox only occasionally, if ever, consume fish, a sharp contrast the feed habits of mink. Their practical nutrition consists mainly of small rodents, hares, birds and their eggs. It is thus unlikely that wild foxes acquired significant levels of omega-3 PUFA, suggesting that fox may not have developed a fatty acid metabolic system capable of oxidizing long chain omega-3 PUFA. With a minor intake of fish, fox in the wild have a genetic basis of intermediary metabolism, which itself is a consequence of evolutionary adaptation to available food resources (Nelson and Ackerman, 1988).

A number of experimental reports indicate a sub-optimal performance of foxes in all phases of the fox ranch year with the employment of dietary regimens employing high levels of omega-3 PUFA. Relative to the reproduction/lactation performance of the fox, studies with blue foxes indicate increased mortality of the females, enhanced number of barren females, and a higher mortality of the pups (Enggaard-Hansen et al., 1991; Rouvinen and Niemela, 1992). In terms of fur development, high levels of fish oil in blue fox diets yielded reduced skin length and impaired guard hair growth (Ahlstrom and Skrede, 1995b).

Requirements—Quantity

Specific guidelines from Scandinavia relative to optimum levels of energy nutrients for fox ranch diets throughout the year are provided in Table 4.16.

TABLE 4.16 Recommended Distribution of Metabolic Energy (%) from Protein, Fat, and Carbohydrates in Fox Diets–Nordic Association of Agricultural Scientists

Period	Protein	Fat	Carbohydrates
December-Whelping	35	20-45	35
Whelping- 7/15	37	35-50	25
7/15 - August	28	35-55	30
September-Pelting	26	35-55	35

* Enggard-Hansen et al., (1991).

Growth/Fur Development

According to studies by Hoie and Rimeslatten (1950) with silver foxes, a high fat/carbohydrate ratio diet provided the most rapid growth, but fur production was superior with diets providing a low fat/carbohydrate ratio.

Reproduction/Lactation

Experimental studies on high fat/carbohydrate ratios relative to the reproduction/lactation performance of the fox are of interest. A high fat/carbohydrate ratio promoted a significantly higher plasma cholesterol and acetoacetate content. Pup mortality was higher in the group with the highest fat/carbohydrate ratio, but pup growth was reduced in the group on the lowest fat/carbohydrate ratio (Ahlstrom, 1992).

These observations of higher pup mortality in litters from females receiving higher fat/carbohydrate ratio dietary regimens had been noted earlier by Rimeslatten (1976a) and Fors et al. (1990). The latter study noted reduced liver glycogen stores in pups littered by females fed a high fat/carbohydrate ratio one week prior to parturition which would give the pups a lower chance of survival during the first 24 hours.

Requirements–Quality

Studies by Rapoport (1961a) with blue foxes indicated the major value of extra vitamin E when employing oil from marine animals in fox diets. Rapoport (1961c) recommended that large quantities of oxidized fat should not be given to potential breeding vixens in the fall months.

Requirements–Essential Fatty Acids

Experimental studies by Ender and Helgebostad (1951a and 1951b) indicate that fox provided salt-water fish or fish products develop a hyperkeratosis which is characterized by an excessive hornification of the epidermis and, later, a marked production of dandruff (scurf) in the fur. The problem was resolved by supplementation of the fox diet with linseed oil, indicating the possibility that fox require both linoleic and linolenic acid which exist at relatively low levels in fish. The authors recommended that foxes should receive 2 to 3 grams of linoleic and linolenic acid/day to prevent hyperkeratosis and dandruff. Rouvinen (1987) found no clear relation between the level of linoleic acid and fur quality, but the incidence of fur defective pelts provided some hints of the positive effects of linoleic acid.

Carbohydrates

Carbohydrates are named on the basis of their empirical formula, CH_2O (ratio of atoms) as illustrated by glucose, $C_6H_{12}O_6$ or $C_6(H_2O)_6$, implying that glucose is a hexahydrate of carbon and giving rise to the name carbohyrate—"carbo" from the Latin and "hydrate" from the Greek. Animals can synthesize the simple sugar glucose from glycogenic amino acids or the glycerol component of fats (triglycerides) in a process termed gluconeogenesis; however, unlike plants, animals cannot synthesize simple sugars from the very basic resources of carbon dioxide and water.

Carbohydrates are represented by monosaccharides (simple sugars) including glucose (blood sugar), mannose and galactose (a component of milk sugar); the disaccharides such as lactose (milk sugar), sucrose (table sugar) and maltose; and the larger polymers including the digestible polymers such as starch and glycogen (animal starch). As well, there are dextrins, smaller polysaccharides obtained via the hydrolysis (breakdown) of starches, and the indigestible fiber resources cellulose, hemicellulose, protopectins, pentosans, and lignins which are useful as bulk for proper consistency of the animal's feces.

The primary role of carbohydrates in animal nutrition is as an energy resource. They are unique in that simple sugars and highly digestible carbohydrates provide an almost immediate energy resource as compared with fats and proteins, which require delayed metabolic processes to yield the key energy resource, ATP (Adenosine Tri-Phosphate).

Like fats, carbohydrates have a "sparing" effect on dietary protein, i.e., optimum levels of both carbohydrates and fats in a fox's diet can lead to minimal protein levels required for top performance. Carbohydrate hydrolysis products as simple sugars can also reduce the fox's requirements for non-essential amino acids in reverse gluconeogenesis metabolism.

In terms of practical nutrition of foxes on ranches, carbohydrates represent one of the most economical energy resources available, second only to fat in terms of kilocalories of ME (Metabolic Energy) for the fox rancher's feed dollar (Leoschke, 1987a, 1987b). Thus, in terms of feed costs per pelt marketed, ranchers are advised to employ the highest levels of fat and quality fortified cereals in their ranch diets throughout the ranch year consistent with the very top performance of their animals

Physiology

The primary role of carbohydrates in fox nutrition is that of an economical energy resource. At the same time, the key role of the fiber content of carbohydrate resources cannot be ignored. Fiber (indigestible carbohydrate polymers) represent a plant's structural components (cellulose and lignins) and cellular adhesives (protopectins and hemicelluloses). (A comical note: "Providing cellulose for the fox is like giving a man a bottle of wine without a cork screw.") Fiber is useful for proper fecal formation as well as of physiological value in encouraging intestinal tone and mobility.

For the most part, with the exception of lactose (milk sugar containing glucose and galactose) and sucrose (table sugar containing glucose and fructose) as well as molasses (providing both glucose and fructose), the final monosaccharide provided to the blood circulation of the fox is glucose. Glucose is an absolutely essential energy resource for human red blood cells and the brain; it also provides oxaloacetate (via private) for the function of the metabolic energy program known as the citric acid cycle. Sub-optimum glucose dietary levels and starvation can lead to ketosis (high levels of ketone bodies in the blood) and fatty degeneration of the liver.

An interesting study by Tallas and White (1988) on glucose utilization in fed and fasted arctic foxes (Alopex lag opus) maintained on a diet similar in composition to food available in the wild provides insights into the glucose physiology of the fox. Fasting (24 hours) glucose was not significantly different from the fed level (134 mg/dl). Fasting was associated with a significant reduction in glucose space, pool size, total entry rate, and irreversible loss which suggested a decline in gluconeogenesis. Glucose recycling was not significantly different between the fed and fasted states. For the arctic fox, the mechanism for defending blood glucose levels during fasting is based on restricting blood glucose to those tissues with high glucose dependency. Based on the nutrient make-up of the diet, about 33% of glucose metabolism was from dietary carbohydrates, 59% from protein, and 8% from glyceride glycerol.

Requirements

It is generally believed that foxes can utilize cereal grain carbohydrates better than mink (Rimeslatten, 1951). He generalized that the percentage of metabolic energy provided by carbohydrates could be 5-10% higher in fox diets relative to mink diets (Rimeslatten, 1976a). This commentary is consistent with the general observation that top performance fox dietary programs provide higher levels of carbohydrates than top performance mink diets.

Vitamins

In terms of animal nutrition, vitamins are distinct from carbohydrates, fats, and proteins inasmuch as they are (a) not energy resources; (b) required in very small amounts, that is, in terms of milligrams and micrograms per day; and (c) co-factors for the multiple enzymes required for animal physiology.

Vitamins are broadly classified as either fat soluble or water soluble. This classification is related to the fact that the fat soluble vitamins (A,D, E, and K) are stored in the liver and in fat depots of the body with minimal opportunity for excretion mechanisms after excessive intake. On the other hand, the water soluble vitamins (multiple B vitamins and vitamin C) have minimal storage depots within the animal body and significant opportunity of daily excretion via the kidneys.

Animals may have (a) an absolute requirement for a specific vitamin; (b) no dietary requirement for the vitamin, as in the case of mink and fox relative to vitamin C; or (c) a physiological environment, as in the case of choline, with significant synthesis of the vitamin from the amino acid methionine, wherein the synthetic processes are simply unable to meet the animal's total requirement for choline.

With fox nutrition management programs, there is the potential for both primary and secondary deficiencies of vitamins. In a primary vitamin deficiency nutritional regimen there are simple sub-optimum levels of the required vitamin. A secondary vitamin deficiency occurs when initially ample quantities of the specific vitamin exist but specific fox feedstuffs in the ranch dietary mixture bring about sub-optimum levels of the required vitamin. Two specific cases illustrate a secondary deficiency of a vitamin.

A biotin deficiency can occur in foxes on a ranch diet with ample biotin levels. However, the addition of raw eggs to the dietary regimen brings about a biotin deficiency as the avidin protein present in raw eggs combines with the biotin to yield a structure unavailable to the digestive processes of the fox.

A thiamine deficiency can occur in foxes on a ranch diet with ample levels of thiamine. The addition of specific fish species such as fresh water carp, smelt, or ocean herring provides a thiaminase enzyme with the capacity to destroy (hydrolyze) all the thiamine present in the ranch diet.

Quality/Quantity Assessment Purified Diets

Purified diets are an ideal program for the study of the vitamin requirements of an animal, as they can be formulated to provide a nutritional regimen with exact knowledge of the vitamin content. Most fox ranchers are familiar with the advertisement "Ivory Soap - 99.4% Pure" which asserts that the soap chemists knew almost the exact composition of their product. Likewise with purified diets, nutrition scientists know essentially 100% of the composition of the diets they employ in their animal nutrition experimental studies.

The high "purity" of these diets is well illustrated in the purified diet employed in mink and fox nutrition studies at the University of Wisconsin. See Tables 4.17, 4.18, and 4.19

TABLE 4.17 Purified Diet Composition—University of Wisconsin, 1945

Nutrient	%
Carbohydrate Resource	
Sucrose - Table Sugar	66
Protein Resource	
Vitamin Test Casein*	19
Fat Resources	
Cottonseed Oil**	8
Cod Liver Oil***	3
Mineral Resource	
Salts IV****	4

Vitamin Resources

B-vitamins via pure crystalline products and fat-soluble vitamins via cod liver oil fortified with vitamin E (alpha tocopherol) and vitamin D-3.

* Vitamin Test Casein is prepared by subjecting casein (dried cottage cheese) to alcohol extraction to remove all the B-vitamins. It is similar to the process of making the morning coffee. This procedure is followed by heat treatment to remove any alcohol residue.

** Obviously containing significant quantities of vitamin E and thus not employed when studying the vitamin E requirements of an animal. In that case a specially distilled lard would be employed.

*** A rich source of fat-soluble vitamins A and D.

**** Phillips and Hart (1935) - crystalline salts - 100% pure.

With the research tool of purified diets, scientists can determine whether a given vitamin is essential for the health and top performance of an animal, The exact quantity of the vitamin required for each phase of the life cycle, and the signs and symptoms indicating a dietary deficiency of a specific vitamin.

The employment of purified diets for the study of the nutritional requirements of fox and mink were initiated at the University of Wisconsin in 1945, under the leadership of Dr. C. A. Elvehjem (Schaefer et al. 1947a) Researchers there considered the possibility that fox and mink as carnivores might have vitamin requirements distinctly different from the common laboratory animals which were herbivores and omnivores. They even considered the possible discovery of a new vitamin, just as Dr. Elvehjem had discovered Niacin in 1935.

TABLE 4.18 Vitamin Content—UW Purified Diet

Vitamin	mg/kg	Vitamin	mg/kg
Thiamine Chloride	2.0	Folic Acid	1.0
Riboflavin	4.0	Biotin	0.25
Pyridoxine HCl	2.0	i-Inositol	250.0
Calcium Pantothenate	15.0	p-Aminobenzoic Acid	500.0
Niacin	40.0	alpha-tocopherol	20.0
Choline	1,000.0	Menadione (synthetic K)	5.0

TABLE 4.19 Mineral Content of the University of Wisconsin and Cornell University Purified Diets for Mink Nutrition Research

Element	UW*	Cornell**
Calcium - %	0.48	0.56
Phosphorus - %	0.28***	0.53***
Sodium - %	0.24	0.26
Potassium - %	0.52	0.58
Chloride - %	0.36	0.40
Magnesium - %	0.036	0.062
Iron - ppm	140.00	150.00
Zinc - ppm	4.4	40.00
Copper - ppm	2.8	20.00
Manganese - ppm	3.1	61.00
Iodine - ppm	22***	4.10***
Molybdenum - ppm		1.50
Selenium - ppm		0.25
Cobalt - ppm		2.00

* Phillips and Hart (1935), **McCarthy et al. (1966)

*** small quantities present in casein component of research diet.

Fat Soluble Vitamins

Vitamin A

Vitamin A is a nutrient termed all-trans retinal with specific roles in the anatomy and physiology of animals. It is required for the integrity of the epithelial tissues and optimum connective and nervous tissue structures. The epithelial cells include the mucous membranes, the cornea, trachea, lungs, kidney, bladder, urinary tract, testicles, vagina, and gastro-intestinal tract. Thus, the following observations on vitamin A deficiency in animals.

Maintenance of Epithelial Tissues

In a vitamin A deficiency, symptoms arise in these cells inasmuch as the vitamin is required for the formation of lipoproteins and/or mucopolysaccharides which are critical constituents of collagen, elastin, cartilage, and other connective tissues. Thus with the defective glycoprotein structures of a vitamin A deficiency, the epithelial cells become keratinized and the resultant dysfunctional keratinization of the mucous membranes of the eyes may lead to "night blindness"and finally permanent blindness (exophthalmia). Loss of epithelial integrity in the intestinal tract can lead to a greater susceptibility to bacterial infections. The loss of epithelial quality in the bladder and urinary tract can lead to bladder stone formation wherein cellular debris can act as a nucleus for the formation of urinary calculi.

Vision

Because vitamin A is a precursor of a component of the visual pigment, rhodopsin, a vitamin A deficiency has effects on the vision of the animal which can lead to permanent blindness.

Bone Growth

Inasmuch as the formation of collagen and cartilage require vitamin A, bone and joint development is defective. Nerve function is compromised as well and leads to a lack of coordination of the animal's legs.

Gene Expression and Reproduction in the Fox

Obvously, animals provided dietary regimens with sub-optimum vitamin A fortification will have relatively poor reproduction, lactation, growth, and fur development (Nenonen et. al., 2003).

Physiology

Vitamin A is absorbed into the mucosal cells of the intestinal wall after emulsification with fats and bile salts in the intestine, forming micelle structures which facilitate absorption. In the mucosal cells, vitamin A is esterified with palmitic acid which enters the blood circulation via the lymphatic system, is removed from the blood by the liver, and stored. It is obvious that optimum absorption of vitamin A and other fat-soluble vitamins (vitamins A, D, E, and K) is facilitated by an emulsification process involving dietary fats. Hence, with fox dietary programs very low in fat, the absorption process for vitamin A would be minimal.

It is generally agreed that the blood levels of vitamin A are homeostatically regulated to a relatively constant level, while the liver level reflects the dietary vitamin A level provided the animal. Hence, plasma vitamin A levels do not reflect an animal's vitamin A nutritional state except in severe deficiencies or toxicity.

Fox appear to have a lesser ability to build up reserves of vitamin A in the liver than mink, (Rimeslatten, 1968). Studies by Ingo et al. (1989) showed that silver fox livers contained 66-2100 IU of vitamin A/gram of dry matter (mean 880 IU/gram dry matter) whereas the blue fox liver content varied from 115-135 IU/gram dry matter.

Requirement

In most animals, vitamin A, all-trans retinal, can be derived from multiple provitamin carotenoids provided in their diets. In terms of the fox, the animals can utilize beta-carotene as a source of vitamin A, but such utilization is poor and not as efficient as vitamin A per se (Coombes et al., 1940; Basset et al.,1946b). On the basis of these observations, when beta-carotene is being used to satisfy the vitamin A requirement of foxes, a conversion factor of 6.0 should be applied to compensate for inefficient utilization. Thus the recommendation of the National Research Council (USA) (Travis et al.,1982) for the optimum vitamin A nutrition of growing foxes is a minimum of 100 IU of vitamin A or 600 IU (360 micrograms) of beta-carotene/kg body wt/day.

Studies by Smith (1941a and 1942b) indicated that a level of 15 IU of Vitamin A/kg body wt/day in the diet of silver fox pups resulted in their exhibiting characteristic nervous symptoms of vitamin A deficiency. Fox pups fed 25-50 IU of vitamin A/kg body wt/day showed growth equal to that of the control group. No appreciable liver storage of vitamin A was noted until the dietary regimen provided 50-100 IU of vitamin A/kg body wt/day. Studies by Coombes et al. (1940) indicated that dietary levels of vitamin A in the range of 1,500 to 3,000 IU/kg dry matter provided satisfactory growth of silver fox pups. However, vitamin A levels of the liver and blood were very low in the fox at pelting. The National Research Council (USA) (Travis et al.,1982) bulletin recommended a level of 2,440 IU vitamin A per kg of dry matter providing 3,700 kcal ME for rapidly growing foxes. The Finnish Fur Breeders Association recommendation is 3,500 IU of vitamin A/kg dry matter (Nenonen et al., 2003). Studies by Bassett et al. (1946b) indicated a minimum requuirement of 200

IU beta-carotene/ kg body wt/day for significant blood levels and liver storage of vitamin A. Bassett et al. (1948) noted that the addition of ascorbic acid to vitamin A–free diets prevented the occurrence of vitamin A deficiency symptoms and provided for higher levels of vitamin A in the blood serum and livers. Very likely the anti-oxidant action of ascorbic acid protected the minimal levels of vitamin A in the vitamin A–free diet.

Resources

Commercial fortified fox cereals generally contain synthetic vitamin A in the form of relatively stable retinal esters including retinal acetate and retinal palmitate. These ester structures are more stable than free retinal which is sensitive to oxidation by peroxides and other structures commonly present in fox fresh/frozen feedstuffs. The stability of these vitamin A esters in cereal mixtures is noted in a study conducted by National Fur Foods Co., (Leoschke, 1957). A commercial fortified mink cereal containing 13, 200 IU vitamin A/kg was placed in a warehouse in January and a sample obtained six months later indicated a vitamin A content of 13, 600 IU/kg. Without question, anti-oxidants including vitamin E are useful for minimizing the loss of vitamin A in fox feedstuffs.

In terms of practical fox nutrition involving acid preserved fish, it is of interest to note that at a pH of 4.5 or lower, partial isomerization of vitamin A from the all-trans isomer to the cis form occurs during storage. This structural change reduces the utilization of the vitamin to about 75%, (De Ritter, 1976; Fog, 1974).

Nutritional Deficiency

Studies on vitamin A deficiency in silver foxes have been conducted by Smith (1941,1942b) and Bassett et al. (1948). A vitamin A deficiency is usually manifested by the arrest of growth and an increased mortality of the pups. With a dietary program deficient in vitamin A, silver fox pups develop a series of nervous derangements which usually begin with a trembling of the head. This is followed shortly by head "cocking," whirling and weaving with a staggering gait, which indicates a sense of balance definitely disturbed. The pups also were noted to run rapidly in a circle, especially when excited, lasting for some 10-15 minutes. A more delayed symptom included a coma lasting 5-15 minutes.

Additional signs included night blindness, exophthalmia (dryness of the eyes), ulceration and rupture of the eyeballs, and papillary edema in the more chronic state of deficiency. Stratification and keratinization of the epithelium of the cornea, trachea branch, kidney, pelvis, urinary bladder, and vagina, as well as widespread myelin degeneration of the spinal cord and cranial nerves also occurred. The cornified epithelial cells do not appear in the vagina until the late stages of the vitamin A deficiency, the nervous symptoms preceeding vaginal cornification by several weeks. Coombes et al. (1940) also noted an inflammation of the mucus membranes of the stomach, intestine, and urinary tract. Often the kidneys and urinary bladder contained calculi. The livers of foxes with vitamin A deficiency symptoms contained no chemically detected quantities of vitamin A. Yet, a nutritional deficiency of vitamin A in foxes has very little effect on the quality of fur development (Picard and Bakker, 1939; Smith, 1942b).

Toxicity

A fox can tolerate large doses of vitamin A, as Helgebosted (1955) notes. A dosage of 40 IU of this vitamin/ gram/ body weight, administered daily over a period of 3 to 4 months produced no toxic signs; however, 200 IU of vitamin A/gram/ body weight/day administered over a period of one to two months produced signs of hypervitaminoses A in fox pups. Signs of toxicity included anorexia, bone changes with exostoses, decalcification and spontaneous fractures, loss of fur, exophthalmia, cramps, and local hyperesthesia of the skin.

Vitamin D

There are two precursors of vitamin D: the plant sterol ergosterol, which can be converted to vitamin D_2 (ergocalciferol) by ultraviolet light provided by sunshine, and the animal sterol present in the skin, 7-dehydrocholesterol, which upon exposure to sunlight is converted to vitamin D_3 (cholecalciferol).

Vitamin D is required for optimum skeletal development in the young, and a deficiency termed rickets can lead to a deformed skeleton.

Physiology

Vitamin D is not physiologically functional until modified in the liver with the addition of a hydroxy group at carbon 25 and later in the kidney where another hydroxy group is added to yield 1, 25-dihydroxycolecalciferol (1, 25-OH D-3), the metabolically active form of vitamin D. From the kidney via the blood circulation, 1, 25-OH D-3 reaches the intestinal mucosa cells wherein it regulates the synthesis of a calcium-binding protein. This protein in turn transports calcium throughout the body via the blood circulation. In addition to 1, 25-OH D-3, the parathyroid hormone (PTH) plays a key role in calcium physiology. The formation of 1, 25-OH D-3 is regulated according to the body's requirement for calcium by the blood serum level, which in turn regulates the release of PTH. A low serum calcium level initiates the release of PTH, which yields an increased formation of calcium-binding protein, enhancing calcium absorption from the intestine.

Requirement

An animal's requirement of vitamin D is determined by multiple factors including the calcium/phosphorus ratio of the diet. Experimental studies of Smith and Barnes (1941) indicated that with an optimum Ca/P ratio of 1/1, rickets was not produced, even with an experimental ration relatively low in vitamin D. However, vitamin D was required for optimum bone growth. A diet providing 0.1% calcium and 0.52% phosphorus, 95 IU of vitamin D/fox/day did provide slight bone healing in one rachitic pup, and 100 to 200 IU of vitamin D/fox/day provided good healing in two fox pups. Studies by Harris et al. (1951b) indicated that a control diet of natural feedstuffs that assayed 0.82 IU of vitamin D/gram or 22 IU per 100 kcal ME was adequate for growing foxes.

Resources

Commercial fox fortified cereals generally contain ample levels of synthetic vitamin D for optimum skeletal development.

Nutritional Deficiency

Rickets is a constitutional nutritional disease of young growing animals. Cases of rickets in the wild are rare but the disease can be produced in young foxes provided diets low in vitamin D in combination with abnormal Ca/P ratios (Hanson, 1935; Schoop, l939; Ott and Coombs, 1941; Smith and Barnes, 1941; Harris et al., 1945; Enders et al., 1949; Helgebostad and Bohler, 1949; Harris et al., 1951b).

Rickets is characterized by a failure of bones to develop in a normal pattern with resultant deformities of the skeleton yielding enlargement of the joints of the front legs which become thickened and crooked as a result of disordered bone trabeculae and osteoid structures. As a result of their bowed condition, the animals appear to be considerably shorter than normal.

Studies of Ender et al. (1949) are of particular interest. With the classical vitamin D deficiency of Steenbock and Black obtained by a high Ca/P ratio, the foxes were quiet with no signs of nervousness. However, with a ricketogenic diet of low calcium with a sufficient or high content of phosphorus, as the blood calcium levels dropped to 6-7 mg % there was tetany with symptoms of spasmophilia, that is, the animals were in a constant state of nervousness and increased reflex irritability.

Toxicity

Studies by Harris et al. (1951b) indicated that at a level of 200 IU of vitamin D/fox/day, there was no toxicity for the young fox. Later experimental work by Helgebostad and Nordstogen (1978) in studies with both silver and blue fox indicated that while the level of 5,000 IU vitamin D/kg body wt/day for two months did not produce clinical symptoms, at a level of 10,000 IU vitamin D/kg body wt/day within a short time the animals showed a loss of appetite and had difficulty in moving, were apathetic, and developed a diarrhea with dark colored feces. Analyses of the blood serum showed a marked Hypercalcaemia with calcium deposits in the kidneys and in some cases in the muscles, gastric mucosa, and the cardiac system. These experiments indicated that fox are relatively resistant to large doses of vitamin D3. However, they might suffer excessive intake of vitamin D through commercial fox ranch diets that may contain high levels of ocean fish where fish liver and fatty tissues are rich sources of vitamin D3 (Jorgensen, 1977).

Vitamin E

Vitamin E, alpha-D-tocopherol, was first discovered in 1922 as a fat-soluble substance in vegetable oils required to maintain pregnancy in rats. Later research studies indicated that vitamin E had significant additional roles in animal nutrition and physiology, including as a factor in the development of muscle dystrophy and as a very important natural anti-oxidant, that is, as a free radical scavenger. As such, vitamin E is part of the cellular defense system protecting the lipid moiety in membrane structures from deleterious peroxidation reactions. Of the eight naturally occurring vitamin E isomers, alpha-D-tocopherol is the most potent in preventing clinical signs of vitamin E deficiency.

Physiology

Peroxides and free radical structures are formed in animal physiology by normal metabolic processes which may be key factors in aging. Vitamin E, in its role as a natural antioxidant, reacts with these peroxide and free radical structures to inactivate them and in the process vitamin E is destroyed. The enzyme glutathione peroxidase, containing the trace element selenium, also functions in the detoxification of peroxide structures.

The trace mineral selenium has a positive interrelationship with vitamin E in animal nutrition and physiology, with the capacity to minimize vitamin E deficiency symptoms and vitamin E requirements. Other trace minerals including copper, iron, and manganese have a negative effect on vitamin E nutrition since they have a pro-oxidative effect on the vitamin E content of feedstuffs, that is, they have the capacity to accelerate vitamin E loss via oxidation.

It is well established that the absorption of alpha-tocopherol from the small intestine of monogastric animals is dependent on multiple factors including pancreatic enzymes, bile salts, and dietary ingredients. Among these are polyunsaturated fatty acids (PUFA) which are very sensitive to peroxidation with resultant potential to destroy vitamin E. Thus, fox ranch diets containing feed resources with high levels of PUFA need higher levels of vitamin E. Adding fish oil to fox diets increases the requirement for vitamin E. With the employment of 12 to 24 grams of fish oil/day to the diet of blue foxes there was noted an adverse affect on sexual activity in males and on fertility in females. A reduced number of pups born unless extra vitamin E was provided (Rapoport, 1961a). The addition of 12-18 grams of fish oil/day to the diet of young blue foxes resulted in increased weight gains; even greater gains were achieved when extra vitamin E was provided (Rapoport, 1961b).

Requirements

Since there are no specific studies reported in the scientific literature on the vitamin E requirement of foxes, these recommendations are based on experimental work with mink. Stowe and Whitehair (1963) noted that 25 mg/kg of dry matter or 6.6 mg/ 1,000 kilocalories of ME was ample for mink. Harris and Embee (1963) recommend an extra level of vitamin E directly related to the level of PUFA present, that is 0.6 mg/gram of PUFA diet.

Resources

The vitamin E resource of choice for modern fox nutrition is stabilized vitamin E, wherein alpha-D-tocopherol is esterified with organic acids such as acetic or palmitic. Leoschke's (1982) studies at the National Research Ranch, indicated that the loss of vitamin E in their mink pellet formulations was only 20% after six months storage.

Nutritional Deficiency

(1) Primary Vitamin E Deficiency

The scientific literature contains no reports on a specific study of vitamin E deficiency in fox that has not been brought about by the presence of high levels of PUFA. In Experimental studies with mink on vitamin E deficient diets without high levels of PUFA, the animals exhibited skeletal and cardiac sympathy with hematologic alterations which included erythrocyte fragility leading to anemia (Stowe and Whitehair, 1963).

(2) Secondary Vitamin E Deficiency

The classic secondary vitamin E deficiency in fur animals known as "yellow fat" disease is brought about by a combination of high levels of PUFA and sub-optimum levels of vitamin E. As early as 1942-1943, Helgebostad and Ender (1955) carried out experiments with silver fox cubs on diets containing herring and round coalfish. At pelting, it was noted a strong yellow color in the adipose tissue which yielded a high iodine value on analysis. The signs of "yellow fat" disease was most evident among the fastest growing male pups with the highest intake of PUFA. Generally, the appetite fails shortly before death. Feces are usually normal, but may frequently become dark as death approaches. The animal becomes listless and dull and its reactions are very slow during the last 24 hours. Urinary incontinence is common.

According to Ender and Helgebostad (1953) and Helgebostad and Ender (1973) additional signs of "yellow fat" disease include the following:

1. Percent of prothrombin and reconvertin is abnormally low in advanced cases of the disease;
2. Anemia is common in chronic cases due to a pronounced fragility of the red blood cells, and there is an increase in the number of leucocytes and thrombocytes;
3. Blood picture reveals aniscytosis and polychromasia;
4. Marked calcification occurs in the endocardium and the endothelium of the large blood vessels of the muscles and kidneys;
5. The spleen is enlarged as much as several times normal size;

6. Extramedullary haematopoiesis is observed;
7. Mucous membranes of the stomach and intestine may be severely thickened with diabetes like bleeding.

Travis and Pilbean (1978) recommend that in addition to vitamin E supplementation of the fox dietary program, fox ranchers employ ethoxyquin (santoquin), a powerful anti-oxidant, in the processing of fish being prepared for fox ranch diets. These synthetic anti-oxidants have been shown to reduce vitamin E loss during fox feed storage and to minimize peroxide development. However, some cautions apply.

Experimental studies by Rouvinen and Laine (1991) on the acute toxic effects of ethoxyquin in the blue fox are of interest. Using supplemental levels of 0, 200, 500 and 1,000 ppm in the diet they found that the highest dosage reduced the appetite dramatically, with subsequent major body weight loss over a 12 day experimental period. Alanine transferase (ALAT) levels were found to be higher in the 500 and 1,000 ppm groups.

Toxicity

The scientific literature on fox nutrition does not provide any data on vitamin E toxicity. However, a field observation of Wilson (1983) with mink kits is of interest. The Wilson observation involved mink on a pellet program providing about 200 mg vitamin E/kg of dry matter. The toxic effect of excessive levels of vitamin E in the absence of high PUFA levels was multiple deaths due to severe hemorrhaging. This effect stemmed from a vitamin E metabolic product antagonistic to vitamin K.

Vitamin K

Vitamin K was named by the Danish scientist, Henrik Dam, who noted that baby chicks on vitamin K free diets had an imparied blood coagulation (koagulation in Danish). Thus, with a vitamin K deficiency, an animal may bleed to death with a minor injury.

Physiology

Vitamin K is known to be an essential co-factor for a carboxylase enzyme that converts glutamyl residues of a number of precursor proteins to Gla (gamma carboxyl-glytamyl) residues required for a variety of coagulation proenzymes for proper function. Inasmuch as vitamin K is a fat-soluble vitamin, it is absorbed along with fats into the body's lymphatic system and finally into the blood circulation. Hence, the absorption of vitamin K requires the presence of bile salts (emulsifying agents) and a healthy functioning intestinal tract.

Requirements

The synthesis of vitamin K by the intestinal flora of animals is a significant and adequate resource for most animals. However, circumstances may arise wherein vitamin K supplementation of the diet is warranted. Perel'dik et al. (1972) reported that on farms where silver and blue foxes were born with subcutaneous and internal organ hemorrhaging, the enrichment of the diet of pregnant foxes with vitamin K was beneficial.

Resources

Synthetic vitamin K (Menadione sodium bisulfite) is the most common resource included in commercial fur animal fortified cereals. There are a number of other products on the market including Heterozyne K and Klotogen K. In Russia, the commercial vitamin K product is called Vicasol.

Nutritional Deficiency

A vitamin K deficiency leads to reduced blood prothrombin levels with subsequent hemorrhages within the body as well as external bleeding.

Toxicity

Perel' dik et al. (1972) noted that a dosage of 6 mg of vitamin K per animal/day yielded symptoms of dyspepsia, nausea, and intensified production of saliva. An intake of 10 mg/animal/day by pregnant mink yielded intoxication within seven days, accompanied by inviable kits.

Water Soluble

Vitamin C–Ascorbic Acid

Scurvy, a vitamin C deficiency in humans, has been known at least since the period of the Crusades. Ascorbic acid is a relatively simple compound synthesized from the common sugars glucose and galactose in most animals. Exceptions include primates (humans and monkeys), guinea pigs, fish, and a number of exotic species. The inability of these species to synthesize ascorbic acid is due to a common defect, the absence of the microsomal enzyme L-gulonolactone oxidase.

Physiology

A deficiency of ascorbic acid impairs a number of physiological activities including the biosynthesis of collagen. With scurvey there is a reduction of the activity of dopamine-beta-hydroxylase and tyrosine hydroxylase (required for melanin pigment synthesis). Also the hydroxylation of lysine and proline is limited.

Requirement

There is no scientific data supporting the concept that fox require supplementary vitamin C in their diets. Fox have the physiological capacity to synthesize ascorbic acid, as numerous experimental studies attest including those of Mathieson (1939, 1942); Lund and Ringsted (1939); Morgan and Simms (1940); Pelletier and Keith (1974); and Helgebostad (1980).) However, all factors considered, there may be a relationship between vitamin A and ascorbic acid in the nutrition of fox, although that basic point remains in question (Bassett et al., 1948). The experimental work of Travis et al. (1982) suggested that with a deficiency of vitamin A in the diet, the synthesis of vitamin C was reduced to below physiological requirements. A deeper insight and understanding of the experimental data indicate that when the basal diet (containing only traces of beta-carotene, a vitamin A precursor) was supplemented with vitamin C, the net result was an intestinal mucosa environment minimizing the destruction of vitamin A formed from beta-carotene. Thus the net result of the vitamin C supplementation was that ascorbic acid acted as an anti-oxidant.

The net result was higher levels of vitamin A in the blood and livers of the foxes. Thus, as Hickman et al. (1944) noted, it appears that vitamin C may have a sparing action on the vitamin A content of the basal diet.

Resources

In terms of practical fox nutrition management, vitamin C (ascorbic acid) is available in a stabilized powder which is useful only in an acidic environment. Thus, within the alkaline (higher pH) environment of most fresh/frozen fox diets, assuming local hard water resources and no phosphoric acid for feed preservation, the vitamin is readily destroyed within a few hours by environmental oxidation.

Toxicity

Vitamin C is relatively non-toxic inasmuch as it is too readily destroyed by the physiological environment and/or eliminated from the body via the kidneys.

Thiamine–Vitamin B_1

Initial studies on the unknown nutritional factors required by animals led to the discovery of a fat-soluble A and a water-soluble B. Further research led to the discovery of multiple members of the B-complex group. The initial discovery relative to the water-soluble B group was vitamin B_1 or thiamine. Thiamine was named for its unique double heterocyclic structure containing a sulfur atom (thio) and a simple primary amine structure.

In fact, the term vitamin in modern animal nutrition is a derivative of "vital amines," unspecified nutritional factors "vital" for the existence of life and containing an amine functional grouping. Later studies, however, indicated this definition did not apply to all vitamin structures.

Physiology

In animal physiology, thiamine is activated by adenosine triphosphate (ATP) to yield thiamine pyrophosphate (TPP). TPP is required for the oxidative decarboxylation of alpha-keto carboxylic acids such as pyruvate, alpha-keto-glutarate, and the keto analogs of leucine, isoleucine, and valine and for the action of the transketolase enzyme of the pentose phosphate pathway of metabolism. Inasmuch as pyruvate is the end product of simple sugar catabolism, the higher the level of carbohydrates (resources for the simple sugars), the higher the requirement for thiamine in an animal's diet.

Requirement

A number of experimental studies have been conducted on the thiamine requirement of the fox, including those of Kringstad and Lunde (1940b), Hodson and Smith (1942a), Ender and Helgebostad (1943), and Harris and Loosli (1949). While these recommendations range from 0.8 mg to 5 mg/kg of dry matter, the National Research Council (USA) (Travis et al., 1982) recommendation is 1.0 mg/kg of dry matter or 27 micrograms/ 100 kilocalories of ME. To protect silver fox females and their offspring from a thiamine deficiency, daily doses of 0.6 to 1.2 mg are recommended by Ender and Helgebostad (1943).

Resources

Cereal grains and brewers yeast are excellent sources of thiamine for fox nutrition, as well as commercial cereals fortified with thiamine.

Nutritional Deficiency

(1) Primary Deficiency

Signs and symptoms of a thiamine deficiency in foxes include anorexia, growth cessation, weakness, cramps, convulsions and paralysis, progressive ataxia, and spastic quadriplegia, with death occurring within 48-72 hours after the onset of neurologic symptoms. The physiological basis for the neurologic symptoms is directly related to the fact that TPP is required for the oxidative decarboxylation of private to yield acetyl Co-A and carbon dioxide. Thus, with a thiamine deficiency, private and its reduction product, lactate, accumulate in the blood and tissues resulting in neurological symptoms. Hyperesthesia is common and the foxes seem to be in great pain as they moan continuously in the advanced stage of the disease. Rectal temperatures taken prior to and during the spastic periods indicated invariably a significant drop in body temperature (Hodson and Smith, 1942a; Coombes, 1940; Green et al., 1941). Studies by Loew and Auston (1975) indicate that fox with a thiamine deficiency had lower erythrocyte transketolase activity, vitamin B_1 being the co-enzyme for that specific enzyme.

In pregnant silver foxes fed diets low in thiamine the net result was abortion or the birth of stillborn pups or if live birth occurred, the mother would eat them. Female fox with a diet deficient in thiamine in the lactation period were observed to bite the tails of all the pups. Some days later appetite decreases; and later the mother will eat some or all of the pups. Classical symptoms of beri-beri (name of thiamine deficiency in humans) appear shortly (Ender and Helgebosted, 1939).

(2) Secondary Deficiency–Thiaminase Fish

The first fox rancher observation of fish induced thiamine deficiency directly related to the presence of a thiamine enzyme occurred on the Chastek Fox Farm, Glencoe, Minnesota in 1932. Hence, the name "Chastek Paralysis" for the spastic paralysis (encephalopathy) wherein the legs are rigidly extended and the head drawn far back (Green, 1938). The disease is characterized pathologically by brain vascular lesions identical to those described by Wernicke in human thiamine deficiency (Green and Evans, 1940; Green et al., 1941; Evans et al.,1942). Experimental studies indicated that the thiaminase inactivation factor was not present in the somatic muscle or fillets of carp (Green et al., 1942a). However, not all fish contain the thiaminse enzyme, as Tables 4.20 and 4.21 indicate.

The problem of thiaminase enzyme in specific fish can be resolved by boiling the fish for a minimum of 15 minutes which denatures the enzyme. Another practical approach to the problem of thiaminase enzyme in fish is to feed the fish on alternate days with extra thiamine supplementation on the non-thiaminase fish days (Green et al; 1942a; Ender and Helgebostad, 1943; and Leoschke, 1969).

Observations of the presence or absence of the thiaminase enzyme in fish, mollusks, and crustaceans is provided in Table 4.20 and Table 4.21.

TABLE 4.20 Occurrence of Thiaminase in Fish, Mollusks, and Crustaceans

Species	Habitat	Source	References
Alewife *(Pomolobos pseudoharengus)*	F	Lake Michigan	Gnaedinger (1965)
Alewife *(Alosu pseudoharengus)*	F	Lake Michigan	Neilands (1947)
Anchovies, striped *(Anchoa hepsetus)*	S	Gulf of Mexico	Jones (1960)
Anchovies *(Engraulis mordax)*	S	Pacific	Stout, et al.(1963)
Bass, white *(Lepibema chrysops)*	F	Great Lakes	Deutsch & Hasler (1943)
Bass, white *(Morone chrysops)*	F	Great Lakes	Deutsch & Hasler (1943)
Black quahog *(Artica islandica)*	S	Atlantic	Lee (1948) Greig & Gnaedinger (1971)
Bjorken *(Abramis blicca)*	F	Lake Malaren	Lieck & Agren (1944)
Bowfin (dogfish) *(Amia calva)*	F	Arkansas	Gnaedinger (1965)
Bream *(Abramis brama)*	F	—	Kuusi (1963)
Buckeye shiner *(Notropus atherionoides)*	S	—	Less (1948) Deutsch & Hasler (1943)
Buffalofish *(Ictiobus prinellus)*	F	Arkansas	Borgstrom (1961)
Bullhead *(Ameirurus m. melas)*	F	Great Lakes	Deutsch & Hasler (1943)
Bullhead *(Ictalurus spp)*	F	Arkansas	Greig & Gnaedinger (1971)
Burbot *(Lota lota maculosa)*	F	Great Lakes	Deutsch & Hasler (1943) Gnaedinger (1965)
Burbot *(Lota lota)*	F	Lake Erie	Deutsch & Hasler (1943) Gnaedinger (1965) Greig & Gnaedinger (1971)
Butterfish *(Poronotus triacanthus)*	S	Gulf of Mexico	Lee et al. (1955)
Capelin *(Mallotus villous)*	F	Arctic	Cehu et al. (1964)
Carp *(Cyprinus carpio)*	F	Great Lakes	Deutsch & Hasler (1943) Gnaedinger (1965)
Catfish, channel *(Ictalurus lacus-tris punctatus)*	F	Great Lakes	Deutsch & Hasler (1943)
Catfish *(Ictalurus nebulous)*	F	Nova Scotia	Nielands (1947)
Chub, creek *(Semotilus a. atromaculatus)*	F	Great Lakes	Deutsch & Hasler (1943)
Clams, chowder, steamer, cherrystone	F	—	Melnick et al. (1945)
Clara *(Mya arenaria)*	S	Atlantic	Neilands (1947)
Crucian *(Cyprinus casarassius)*	F	Sweden	Lieck & Agren (1944)
Fathead minnow *(Primephales p. promelas)*	F	Great Lakes	Deutsch & Hasler (1945)

Species	Habitat	Source	References
Garfish, garpike	S	—	Borgstrom (1961)
Garfish *(Belone acus)*	F	Sweden	Lieck & Agren (1944)
Goldfish *(Carassius auratus)*	F	Great Lakes	Deutsch & Hasler (1945) Gnaedinger (1965)
Herring, Baltic *(Clupea harengus var. membranus)*	S	Baltic	Kuusi (1963)
Herring *(Clupea harengus)*	S	Atlantic	Deutsch & Hasler (1945) Nielands (1947)
Ide *(Leuciscus idus)*	F	Lake Malaren	Lieck & Agren (1945)
Lamprey eel, adult *(Petromyzon marinus)*	F	Great Lakes	Borgstrom (1961)
Lobster *(Homarus americanus)*	S	Atlantic	Nielands (1947)
Mackerel, Pacific *(Scomber japonicas)*	S	Pacific	Borgstrom (1961)
Menhaden *(Brevoortia tyrannus)*	S	Chesapeake Bay	Greig & Gnaedinger (1971)
Menhaden, large scale *(Brevoortia patronus)*	S	Gulf of Mexico	Jones (1960)
Menomonee whitefish *(Prosopium cylindrac-em quadrilaterale)*	F	—	Deutsch & Hasler (1945)
Moray eel *(Gymnothorax ocellatus)*	S	Gulf of Mexico	Lee et al. (1955)
Minnow, mud *(Umbra limi)*	F	Great Lakes	Deutsch & Hasler (1945)
Minnow *(Fundalus heteroclites)*	F	Nova Scotia	Nielands (1947)
Minnow, fathead *(Pimephales p. promelas)*	F	Great Lakes	Deutsch & Hasler (1943)
Minnow *(Fundalus diaphanus)*	F	Nova Scotia	Nielands (1947)
Mussel *(Elliptio complanatus)*	F	Nova Scotia	Nielands (1947)
Mussel, bigtoe *(Pleurobema cordatum)*	F	Tennessee River	Gnaedinger (1965)
Mussel *(Mytilus edulis)*	S	Atlantic	Nielands (1947)
Quahogs *(Venos mercenaria)*	S	—	Soc. Exp. Bio.1942 60: 268-269
Razor belly, scaled sardine *(Harengula Pensacolae)*	S	Gulf of Mexico	Lee et al. (1955)
Rudd *(Leuciscus erythrphtalmus)*	F	Lake Malaren	Lieck & Argen (1944)
Sauger pike *(Stizostedion c. canadense)*	F	Great Lakes	Deutsch & Hasler (1943)
Trench *(Tinca vulgaris)*	F	Lake Dreugen	Lieck & Argen (1944)
Vae *(Ahiamas vinabra)*	F	Great Lakes	Lieck & Argen (1944)
White bass *(Lepimbema chrysops)*	F	Great Lakes	Deutsch & Hasler (1943)
Whitefish *(Prosopium cylindraceum quadriaterale)*	F	Great Lakes	Deutsch & Hasler (1943)
Whitefish *(Coregonus clupeaformis)*	F	Great Lakes	Deutsch & Hasler (1943) Bell & Thompson (1951)

TABLE 4.21 Fish, Mollusks, and Crusteans without Thiaminase Enzyme

Species	Habitat	Source	References
Ayu *(Plecoglossus altivelis)*	F	—	Borgstrom (1961)
Bass, small mouth *(Micropterus dolomieu)*	F	Great Lakes	Deutsch & Hasler (1943)
Bass, large mouth *(Huro psalmodies)*	F	Great Lakes	Deutsch & Hasler (1943)
Bass, rock *(Ambloplites r. rupestris)*	F	—	Deutsch & Hasler (1943)
Black backs *(Pseudopleuronectes Americanus)*	S	Atlantic	Neilands (1947)
Bluegill *(Lepomis m. marchers)*	F	Great Lakes	Deutsch & Hasler (1943) Gnaedinger (1965)
Chub, bloater *(Coregonus hoyi)*	F	Lake Michigan	Deutsch & Hasler (1943) Gnaedinger (1965)
Cod *(Gadus morrhua)*	S	Atlantic	Deutsch & Hasler (1943)
Crappie *(Pomoxis nigro-maculates)*	F	Great Lakes	Deutsch & Hasler (1943)
Croaker *(Micropogon undulates)*	S	Gulf of Mexico	Lee et al. (1955) Gnaedinger (1965)
Cunner *(Tautogolabrus adsperus)*	S	Long Island Sound	Lee (1948) Soc. Biol. (1945)53:63
Cusk *(Brosme bromse)*	S	Atlantic	Neilands (1947)
Cutlassfish, silver eel *(Trichiurus lectures)*	S	Gulf of Mexico	Lee et al. (1955) Gnaedinger (1965)
Dogfish *(Squalus acanthias)*	S	Atlantic	Neilands (1947)
Dogfish *(Amia calva)*	—	—	Deutsch & Hasler (1943)
Eel *(Anguilla rostrata)*	F	-----	Neiland (1947)
Eelpout *(Zoarces aguillaris)*	S	Atlantic	Neilands (1947)
Garpike, Northern longnose *(Lepisosteus osse-us oxyurus)*	F	—	Deutsch & Hasler (1943)
Goosefish *(Lophins piscatorial)*	S	—	Neilands (1947)
Haddock *(Melanogrammus aeglefinus)*	S	Atlantic	Deutsch & Hasler (1943)
Halibut *(Hippoglossus hippoglossus)*	S	Atlantic	Neilands (1947)
Hake, Pacific *(Merluccius productis)*	S	Pacific	Stout et al. (1963)
Hake, silver *(Merluccius bilinear is)*	S	Atlantic	Rouvinen et al. (1997)
Hake *(Urophycis spp.)*	S	Pacific Gulf of Mexico	Stout et al. 1963) Lee et al (1955)
Herring *(Leucichthys artedi areturus)*	F	Lake Superior	Deutsch & Hasler (1943)
Jackfish *(Esox lucius)*	F	—	Bell & Thompson (1951)
King whiting (ground mullet) *(Menticirrhus americanus)*	S	Gulf of Mexico	Gnaedinger (1965)

Species	Habitat	Source	References
Lemon sole (*Pseudopleuronectes americanus dignabilis*)	S	—	Deutsch & Hasler (1943)
Ling (*Lota lota masculosa*)	F	—	Bell & Thompson (1951)
Lizard fish (*Synodus foetens*)	S	Gulf of Mexico	Lee et al. (1955)
Lobster (*Homarus americanus*)	S	Atlantic	Neilands (1947)
Lumpfish (*Cyclopterus lumpus*)	S	Atlantic	Greig & Gnaedinger (1951)
Mackerel (*Scomber scombrus*)	S	Atlantic	Deutsch & Hasler (1943)
Mullet (*Mugil spp.*)	S	Gulf of Mexico	Gnaedinger (1965)
Oyster (*Ostrea edulis*)	S	Atlantic	Greig & Gnaedinger (1971)
Perch, white (*Morone Americana*)	F	—	Neilands (1947)
Perch, yellow (*Perca flavescent*)	F	—	Deutsch & Hasler (1943)
Periwinkle (*Littorina litorea*)	S	Atlantic	Greig and Gnaedinger (1971)
Pikerel, Northern (*Esox lucius*)	F	Great Lakes	Deutsch & Hasler (1943)
Pike, walleye (*Stizostedion citreum*)	F	Great Lakes	Deutsch & Hasler (1943)
Plaice, Canadian (*Hippoglossoides platessoides*)	S	Atlantic	Neilands (1947)
Pollock (*Pollachius virens*)	S	Atlantic	Neilands (1947)
Porgy, scup (*Stenotomus aculeate*)	S	Gulf of Mexico	Lee et al. (1955)
Porgy, scup (*Stenotomus chrysops*)	S	Chesapeake Bay	Greig & Gnaedinger (1971)
Pumpkinseed (*Lepomis gibbous*)	F	Great Lakes	Deutsch & Hasler (1943)
Redfish (*Sebastes marinus*)	S		Deutsch & Hasler (1943)
Sculpin (*Myxocephalus Octodecemspinosus*)	S	Atlantic	Greig & Gnaedinger (1971)
Salmon, Atlantic (*Salmo salar*)	F	Atlantic	Neilands (1947)
Salmon, coho (*Oncorhynchus kisutch*)	F	Lake Michigan	Borgstrom (1961)
Seabass (*Centropristis striatas*)	S	Chesapeake Bay	Borgstrom (1961)
Sea catfish (*Arius felis*)	S	Gulf of Mexico	Lee et al. (1955)
Sea catfish (*Galeichthys felis*)	S	Gulf of Mexico	Lee et al. (1955)
Sea raven (*Hemitripterus americanus*)	S	Atlantic	Grieg & Gnaedinger (1971)
Sea robin (*Prionotus spp.*)	S	Gulf of Mexico	Lee et al. (1955) Gnaedinger (1965)
Sea robin (*Prionotus carolinus*)	S	—	Soc Exp Biol (1945) 60:268
Sheepshead, freshwater drum (*Aplodinotus grunniens*)	F	Lake Erie	Grieg & Gnaedinger (1971)
Shrimp, brine (*Artemia salina*)	S	Lab Grown	Grieg & Gnaedinger (1971)
Skate (*Raja senta*)	S	Atlantic	Grieg & Gnaedinger (1971)
Smelt, pond (*Hypomesus olidus*)	F	—	Borgstrom (1961)

Species	Habitat	Source	References
Spot *(Leiostomus xanthurus)*	S	Gulf of Mexico	Lee et al. (1955) Gnaedinger (1965)
Squid *(Loligo brevis)*	S	Gulf of Mexico	Lee et al. (1955)
Starfish *(Asterias vulgaris)*	S	Atlantic	Grieg & Gnaedinger (1971)
Tautog, blackfish *(Tauloga onitis)*	S	Long Island Sound	Lee (1948)
Trout, brown *(Salmo trutta fario)*	F	Great Lakes	Deutsch &Hasler (1943)
Trout, lake *(Christivomer n. namaycush)*	F	Great Lakes	Deutsch & Hasler (1943)
Trout, rainbow *(Salmo gairderii iridous)*	F	Great Lakes	Deutsch & Hasler (1943)
Tullibee *(Leucichthys tullibee)*	F	—	Bell & Thompson (1951)
White trout *(Cynoscion nothus)*	S	Gulf of Mexico	Gnaedinger (1965)
White trout *(Cynoscion aviaries)*	S	Gulf of Mexico	Lee et al. (1955)
Whiting *(Merluccius bilinear is)*	S	Atlantic	Deutsch & Hasler (1943)
Witch flounder *(Glyptocephalus cynogossus)*	S	Atlantic	Greig & Gnaedinger (1971)
Yellow tails *(Lamanda ferruginous)*	S	Atlantic	Deutsch & Hasler (1943)

Riboflavin–Vitamin B_2

Riboflavin is termed vitamin B_2 inasmuch as it was the second water-soluble vitamin to be discovered and identified. The structure of the vitamin consists of an isoalloxazine ring attached to a ribityl side chain. The ribityl structure is similar to the simple pentose sugar ribose. The name riboflavin is derived from its ribityl side chain and the fact that it is a key component of the flavin mononucleotide (FMN) and flavin adenine dinucleotide (FAD) co-enzymes involved in the metabolism of amino acids, fatty acids, and the energy yielding citric acid cycle.

Physiology

In animal physiology, riboflavin is activated by ATP to yield flavin mononucleotide (FMN). In a second biosynthetic step, FMN is combined with a second molecule of ATP to form flavin adenine nucleotide (FAD). Within the citric acid cycle, FAD is involved in the conversion of succinate to trans-fumarate and also in fatty acid and amino acid catabolism. In fatty acid catabolism, FMN is involved in the dehydrogenation of alkane structures to trans-alkene functional groupings. Inasmuch as FAD and FMN are involved in multiple steps of fatty acid catabolism, it is understandable that the riboflavin requirement of an animal increases with the employment of higher levels of fat in dietary programs.

Riboflavin is water soluble and hence excess intake is mainly excreted via the urine (Jorgensen et al., 1975).

Requirements

Experimental work by Schaefer et al. (1947a) with silver foxes indicated that the riboflavin requirement of the fox is greater than 1.25 mg/kg and less than 4.0 mg/kg of diet. The National Research Council (USA) (Harris et al., 1953) on the basis of the Schaefer et al. (1947a) studies recommended 2.0 mg/kg of diet.

In experimental studies with blue foxes, Rimeslatten (1958) indicated that riboflavin deficiency signs were noted with blue foxes on a basal diet calculated to contain 1.3 to 1.6 mg/kg of diet. Rimeslatten recommended that the riboflavin requirement of the blue fox was a minimum of 0.1 mg/100 kilocalories ME after weaning, equivalent to 3.6 mg/kg of dry matter. For blue fox during pregnancy and lactation he recommended a minimum of 0.15 mg/100 kilocalories ME. In a later study, Rimeslatten (1959), the recommendation was an allowance for both blue and silver foxes of 0.4 mg/ 100 kilocalories during pregnancy and lactation and 0.25 mg/100 kilocalories for older cubs.

Resources

Brewers yeast, liver, and kidneys are excellent sources of riboflavin while cereal grains are good sources. Quality commercial fortified cereals provide ample levels of riboflavin.

Nutritional Deficiency

Signs of a riboflavin deficiency in foxes as noted by Schaefer et al. (1947a) included weight loss, muscular weakness, clonic spasms and coma, and opacity of the lens. Diminution of pigment in the underfur and guard hair also occurred.

Niacin

Niacin as a specific chemical was known long before its animal nutrition role was discovered by Elvehjem at the University of Wisconsin in 1935. It was first isolated as an oxidation product of the natural alkaloid, nicotine, from which its name is derived. Niacin became the preferred name rather than nicotinic acid, in deference to the public health concerns about the negative value of the nicotine content of tobacco for one's health.

Physiology

Niacin is a component of two enzymes of metabolism, namely NAD (Nicotinamide Adenine Dinucleotide) and NADP (Nicotinamide Adenine Dinucleotide Phosphate). NAD is a co-enzyme for a number of dehydrogenases participating in the catabolism of fatty acids, simple sugars, and amino acids. NADP also participates in dehydrogenation reactions, particularly in the hexose monophosphate shunt of glucose metabolism. Reduced NADP, that is NADPH, has an important role in the synthesis of fatty acids and steroids.

Requirement

Schaefer et al. (1947a) recommended a minimum requirement is 0.4 to 2.0 mg/kg body weight. The National Research Council recommendation (Travis et al., 1982) is 10 mg/kg of dry diet or 0.26 mg/100 kcal of M.E. Fox ranch diets generally contain 50-70 mg/kg dry matter (Utne, 1974).

Resources

Brewers yeast, baking yeast, and liver are excellent sources of niacin while cereal grains are good sources. Quality commercial fox fortified cereals provide ample levels of niacin.

Nutritional Deficiency

A deficiency of niacin in foxes leads to "Black Tongue" disease. Its typical signs include anorexia, weight loss, severe inflammation of the gums and fiery redness of the lips and tongue, palatine redness, gastrointestinal hemorrhage, and diarrhea according to Hodson and Loosli (1942) and Schaefer et al. (1947a).

Pyridoxine, Pyridoxamine, Pyridoxal–B_6

Vitamin B_6 exists in three interconvertible forms: pyridoxine, pyridoxal, and pyridoxamine. Of these, pyridoxine is the most commonly used as a supplement in animal feeding programs. Although vitamin B_6 is relatively heat stable, various thermal reactions reduce the content of biologically available B_6 in processed foods. Thus, considerable amounts of B_6, especially pyridoxal and pyridoxal phosphate with free aldehyde functional groupings may react with free amino functional groupings of amino acids and proteins during heat treatment to yield less biologically active pyridoxamino structures (Gregory and Kirk, 1981).

Physiology

The physiologically active form of vitamin B_6 is the phosphorylated form brought about by ATP and pyridoxal kinas.

Requirement

A level of 2.0 mg/kg or 50 micrograms/100 kcal of ME of pyridoxine will prevent signs of deficiency in foxes (Schaefer et al., 1947).

Resources

Brewers yeast and baking yeast are excellent sources of vitamin B_6. Quality commercial fortified fox cereals provide ample levels of vitamin B_6.

Nutritional Deficiency

A deficiency of vitamin B_6 in foxes leads to anorexia, cessation of growth, low hemoglobin levels, and a comatose state until death (Schaefer et al., 1947a).

Toxicity

Inasmuch as vitamin B_6 is water soluble, with excessive levels being excreted in the urine, the possibility of a vitamin B_6 toxicity is remote.

Pantothenic Acid—"Anti-Gray Hair" Vitamin

Pantothenic acid is essential for all living organisms and is widely distributed in nature. Because of initial observations of experimental studies with laboratory rats placed on purified diets without pantothenic acid, the vitamin for many years has been referred to as the "anti-gray hair" vitamin.

Physiology

Pantothenic acid is a key component of co-enzyme A (Co-A) which functions as an acetyl group transfer co-factor for many enzymatic reactions in the metabolism of animals and plants. Co-A is involved in acetylating amines, the oxidative-decarboxylation of pyruvate (the end point of the glycolysis pathway of the catabolism of simple sugars), the synthesis of citrate from acetyl Co-A and oxaloacetate, the oxidative decarboxylation of alpha-keto-glutarate within the citric acid cycle, and with the beta-oxidation step of fatty acid catabolism.

Requirement

Experimental work by Schaefer et al. (1947a) employing purified diets indicated that the fox requirement for pantothenic acid was greater than 2.5 mg/kg of dry matter and less than 15 mg/kg of dry matter. The National Research Council (U.S.A.) (Travis et al., 1982) has recommended 8.0 mg/kg of dry diet or 0.21 mg/kilocalorie of ME.

Resources

Yeasts, liver, and skimmed milk powder are excellent resources of pantothenic acid while cereal grains and by-products are good resources for this vitamin. Quality commercial fortified cereals provide ample levels of pantothenic acid.

Nutritional Deficiency

Studies by Lund and Kringsted (1939), Morgan and Simms (1940), and Schaefer et al. (1947a) have provided the following signs of a pantothenic acid deficiency in foxes including weight loss, clonic spasms followed by coma, catarrhal gastroenteritis, cloudy swelling, and congestion of the kidneys.

In terms of fur development, the fox exhibited extensive depigmentation with graying beginning at the snouts and working backwards toward the eye sockets and ears. There was a rapid loss of fur so that their coats appeared short and cottony like that of young animals.

Biotin

As is in the case of most animals, biotin may not be required as a dietary component for foxes inasmuch as intestinal flora synthesis of biotin may be sufficient to meet the animal's requirements.

Physiology

In animal metabolism, biotin is a co-enzyme for a number of acetyl-Co-A carboxylases and for the alpha-keto-carboxylase involved in the conversion of pyruvate to oxaloacetate initiation of the citric acid cycle and for the decarboxylase involved in the conversion of oxaloacetate to pyruvate. A biotin requiring pyruvate carboxylase is also involved in a regulatory pathway in gluconeogenesis. A biotin requiring Co-A carboxylase catalyses, an essential step in fatty acid biosynthesis. Biotin is also a co-enzyme for a number of enzymes involved in the catabolism of amino acids.

Requirement

No experimental data has been reported on the biotin requirement of the fox.

Resources

Excellent sources of biotin include brewers and torula yeast, as well as beef and pork liver. Good sources include cereal grains and their by-products. Quality commercial cereals provide ample levels of biotin for normal fox nutrition, provided that the ranch diet does not contain raw eggs.

Nutritional Deficiency

(1) Primary Biotin Deficiency

All factors considered, it is practically impossible to experience a biotin deficiency in foxes on common fox ranch dietary programs. Both the biotin content of fox feed ingredients and synthesis of the B-vitamin by the animal's intestinal flora make a deficiency unlikely.

(2) Secondary Biotin Deficiency

Fox ranch diets with ample levels of biotin can, however, yield a secondary biotin deficiency if raw eggs or raw egg whites are included in the fox dietary regimen (Ender and Helgebostad, 1958, 1959). Raw eggs contain a protein termed avidin which binds biotin in a structure unavailable to the digestive processes of animals. With the heating of egg products at boiling temperatures for 15 minutes, the protein is denatured and no longer functions as an anti-biotin nutritional factor.

With a secondary biotin deficiency, the foxes exhibited typical symptoms of a biotin deficiency including achromotrichia, "spectacle" eyes, poor fur quality, moulting, and a dermatosis characterized by dryness and roughness of the skin and hyperkeratosis. Fur and tail biting were common symptoms. Animals on a control diet providing boiled egg whites exhibited normal health and fur development.

Choline

Choline was first isolated from hog bile and thus the origin of its name from the Greek word for bile, *chole*. The term lipotropic agent applied to choline refers to its action as a dietary substance that decreases the rate of deposition of abnormal amounts of lipid in the liver and that accelerates the removal of excess fat from it. The term neurotransmitter is applied to the acetyl derivative of choline to designate its action as the chemical mediator of synapses of the autonomic nervous system. Nerve fibers utilizing acetylcholine as a mediator are referred to as cholinergic. In pigs, rats, and mice, choline is required for the prevention of fatty liver and fatty kidneys.

Choline is not a vitamin but a nutrient that is more or less essential in specific animals depending upon the degree of biosynthesis from the amino acids methionine and betaine (methyl group donors) and serine (ethanol amine resource for choline synthesis via decarboxylation).

Physiology

Lecithin (phosphatidyl choline) and choline containing sphingomyelin are key components of biological membranes as well as lipoproteins. The neurotransmitter acetylcholine is synthesized from choline and acetyl Co-A via an enzyme termed choline acetylase. The breakdown (hydrolysis) of acetylcholine is via acetylcholine esterase. Choline is activated via cytosine tri-phosphate (CTP) to yield CDP-choline, the precursor of lecithin and sphingomyelin. Choline is part of a labile methyl pool capable of contributing methyl groups for the synthesis of the amino acid methionine and other methylated compounds including purine and pyramidine, bases necessary for proper growth and cell function.

Requirement

Although choline can be synthesized by animals via available labile methyl group donation from methionine or betaine, according to Perel'dik et al. (1972) fox are unable to meet their choline requirement by biosynthesis alone. The choline requirement of the fox has not been established. Thus Perel'dik et al. (1972) is based on an analogy to dog nutritional requirements and recommended 50-70 mg/kg body weight as a medicinal dosage. In experimental studies with fox [employing purified diets] at the University of Wisconsin, Schaefer et al. (1947) found that a level of one gram/kg proved satisfactory for the growth of silver foxes.

Inasmuch as methionine is a potential source of labile methyl groups. choline can spare the methionine requirement of the fox and vice versa, that is, higher levels of methionine in the fox ranch diet can reduce the fox's requirement for choline. Likewise, betaine with labile methyl groups can reduce the animal's requirement for choline.

Vitamin B_{12} is involved in the biosynthesis of methionine. Thus the level of vitamin B_{12} in the fox diet could have a significant effect on the choline requirement of foxes.

Resources

The richest source of choline is egg yolk. Other excellent sources of choline for practical fox nutrition include animal liver and brain as well as yeasts. Quality commercial fortified cereals provide ample levels of the substance.

Nutritional Deficiency

Without question, the classic sign of a choline deficiency is fatty livers, or hepatic fat infiltration, which can be also brought about via starvation or by animals being "off feed" for a few days.

Folic Acid

Folic acid and vitamin B_{12}, as well as the trace minerals iron and copper, are critical micronutrients required for the prevention of anemia. A deficiency of folic acid in animals yields a characteristic macrocytic, hypochromic anemia.

Physiology

Folic acid is a key vitamin for metabolic function as a carrier of one-carbon units as derivatives of formate, formaldehyde, or methanol. These one-carbon units are generated primarily during amino acid metabolism and are used in the interconversions of amino acids and in the biosynthesis of the purine and pyramidine components of nucleic acids required for cell division.

Requirement

Although folic acid is synthesized by the fox intestinal flora, the level of microbial folate contributed to the physiology is not sufficient to meet the needs of the animal for survival (Tove et al., 1949). This experimental work also indicated that folic acid conjugates including heptaglutamate and pteroylglutamate were incapable of replacing folic acid in the diet of foxes on experimental purified diets.

Schaefer et al. (1947b) suggested that 0.2 mg/kg of dry diet or 5.2 micrograms per 100 kcal ME be accepted as the tentative requirement. Recent studies by Polonen, et al. (2000) indicate that foxes fed a high proportion of silage containing formic acid should receive a supplement of folic acid in excess of 5 mg/kg dry matter during the growth and furring periods.

Resources

The richest sources of folic acid are liver and yeast products. Quality commercial fortified cereals provide ample levels of folic acid.

Nutritional Deficiency

Foxes fed a purified diet deficient in folic acid develop anorexia, loss of body weight, and a decrease in hemoglobin and in red and white blood cells (Schaefer et al., 1947b).

Vitamin B_{12}—Cobalamine

Vitamin B_{12} and folic acid, as well as the trace minerals iron and copper, are critical micronutrients required for the prevention of anemia. A deficiency of vitamin B_{12} in humans yields pernicious anemia. Vitamin B_{12} has a pink color related to the presence of the trace mineral cobalt, hence the extra nomenclature of "cobalamine."

Physiology

Vitamin B_{12} is an absolute requirement for the function of folic acid in animal nutrition, inasmuch as vitamin B_{12} is required for folate uptake by cells and for regenerating tetrahydrofolate (THF) from which is made 5,10 methylene THF required for thymidylate synthesis, a step of metabolism absolutely required for cell synthesis. Vitamin B_{12} is also critical for methyl group transfer including the methyl group of methionine.

Requirement

At the present time, no experimental studies have been reported in the scientific literature on the vitamin B_{12} requirement of foxes. Practical fox ranch diets usually contain ample quantities of vitamin B_{12}, a component of animal protein resources. In addition, there may be significant synthesis of vitamin B_{12} by the intestinal flora of the fox.

Resources

The richest sources of vitamin B_{12}, once known as the Animal Protein Factor (APF), are liver, egg yolk, and whole fish. Milk is a good source, but all plant products lack this specific vitamin.

Nutritional Deficiency

No reports exist in the scientific literature of a specific study on vitamin B_{12} deficiency in foxes. Studies by Schaefer et al. (1948) and Tove et al. (1949) with young foxes on purified diets providing all the the known vitamins including folic acid indicated that fox require an unknown methanol soluble fraction of liver for normal hemoglobin levels. Later experimental studies by Leoschke (1952) with mink indicated that the unknown methanol-soluble liver factor required by both mink and fox was vitamin B_{12}.

Minerals

General Commentary

Minerals have multiple functions in the anatomy and physiology of foxes, including the following:

1. Structural role in the skeleton, teeth, and soft tissues; blood composition and the blood clotting mechanisms, as well as nervous system components;
2. Regulatory role in maintenance of osmotic pressure and acid-base balance;
3. Essential role as co-factors for the activity of many enzymes;
4. Role as vitamin structure, as illustrated by cobalt in vitamin B_{12} and sulfur in biotin;
5. Key role in fur pigmentation, as iron deficiency results in "cotton" mink.

In terms of animal nutrition, the mineral requirements can be divided into two groups. The first group consists of the macro minerals–calcium, chloride, magnesium, phosphorus as phosphate anions, potassium, sodium, and sulfur both as part of the structure of vitamins and amino acids and as sulfate anion. These are required in relatively large quantities. The second group consists of the micro minerals, or trace elements including copper, iron, iodine, manganese, molybdenum, selenium, and zinc. These are needed in relatively small quantities, often expressed in parts per million (ppm).

Absorption of minerals from the gastro-intestinal tract of animals involves multiple factors including (a) relative solubility in the specific environmental pH; (b) length of time of feed passage; (c) presence of factors that interfere with absorption such as conalbumin in egg white and DMNA or formaldehyde presence in certain fish which reduces iron absorption; and (d) the presence of specific vitamins such as vitamin D required for the absorption of calcium and phosphate ions or vitamin B_{12} and ascorbic acid which enhance iron absorption.

In terms of modern fox nutrition, employing chelated minerals may lessen the animal's chance of developing a secondary deficiency of specific trace minerals such as copper, iron, or zinc. The word chelate orginates from the Greek word for claw–*chele*; thus, chelated minerals are metallic cations combined with specific chelating agents such as amino acids to yield a ring-like structure. Chelation shields the mineral from external influences and thereby affects both intestinal absorption and interference from other minerals on absorption. For example, calcium may bind zinc, decreasing its absorption; however, chelated zinc is protected from binding by calcium.

Mineral Imbalances

Calcium/Phosphate Ratios

High or low Ca/P ratios may undercut the absorption of calcium cations or phosphate anions and thus bring about a secondary mineral deficiency.

Calcium/Phosphate/Magnesium Ratios

Some antagonism is recognized among magnesium, calcium, and phosphate; thus, excessive calcium or phosphate in the diet may decrease the absorption of magnesium or vice versa.

Copper/Zinc Ratios

Van Campen and Scaife (1967) and Van Campen (1969) showed that copper and zinc compete for absorption in the intestine of rats and that an excess of either element reduces the absorption of the other.

Molybdenum/Copper Ratios

High levels of molybdenum can induce a copper deficiency. In contrast, low levels of forage molybdenum can be associated with copper toxicity. Arrington and Davis (1953) showed that a high level of molybdenum induced a copper deficiency in rabbits, with typical signs such as anemia.

In terms of practical fox nutrition, ranchers should be careful not to provide excessive ash levels in the diet. Ranch diets with 35- 40% bone-in feed products provide about 7-8% ash (dry matter basis), which contains ample levels of calcium and phosphate. Excessive levels of ash can undermine the nutrition of the fox by creating (1) Excessive bulk, especially harmful during the critical nutritional stress periods of lactation and the early growth of the pups; (2) Excessive ash levels, especially of limestone (calcium carbonate), which can undercut fatty acid absorption via the formation of insoluble calcium soaps in the intestinal tract;(3) Bladder stones (urinary calculi) and/or kidney stones.

As noted earlier in the case of the vitamin nutrition of the fox, the sub-optimum mineral nutrition status of the fox can be classified as primary or secondary. In a primary deficiency, sub-optimum levels of the specific mineral are provided in the dietary program. In a secondary deficiency, ample quantities of the specific mineral are provided in the dietary program but certain specific factors are responsible for binding the mineral in a structure unavailable to the digestive processes of the animal. A good example is the case in which the conalbumin protein in egg white binds iron.

Mineral Requirements

Inasmuch as minimal experimental data exists on the mineral requirements of foxes, fox nutrition scientists should study the mineral levels present in the experimental purified diets employed at the University of Wisconsin and Cornell University in studying the nutritional requirements of mink. See Table 4.22.

TABLE 4.22 Mineral Content of the University of Wisconsin and Cornell University Purified Diets for Mink Nutrition Research

Element	UW*	Cornell**
Calcium-%	0.48	0.56
Phosphorus-%	0.28***	0.53***
Sodium-%	0.24	0.26
Potassium-%	0.52	0.58
Chloride-%	0.36	0.40
Magnesium-%	0.036	0.062
Iron-ppm	140.00	150.00
Zinc-ppm	4.40	40.00
Copper-ppm	2.80	20.00
Manganese-ppm	3.10	61.00
Iodine-ppm	22.00***	4.10***
Molybdenum-ppm		1.50
Selenium-ppm		0.25
Cobalt-ppm		2.00

* Phillips and Hart (1935), ** McCarthy et. al. (1966)

*** small quantities present in casein component of research diet

Calcium/Phosphate

Since calcium and phosphorus are so closely interrelated both nutritionally and physiologically, it makes common sense to discuss them together. Calcium and phosphorus in the form of hydroxyapatite crystals ($_3Ca(PO_4)_2$: $Ca(OH)_2$) are the major minerals involved in the structural rigidity of bones and teeth. Calcium is also involved in blood clotting and in impulse transmission at the neuromuscular junction. Phosphorus is involved in bone and teeth formation and, in addition, is an important structural element for the soft tissues including membrane phospholipid components (lecithin and cephalin), nucleic acids, and proteins as illustrated by the phosphorylated protein of milk, casein, and the vital role of ATP (Adenosine Tri-Phosphate) in energy transfer in animal physiology.

Physiology

Studies by Valaja et al. (2000) with blue foxes involved mineral levels ranging from 5 to 11% ash and Ca/P ratios of 1.2/1 to 1.7/1 via the employment of meat and bone meal and the use of limestone and monocalcium phosphate. The experimental data indicated that both the mineral level and the Ca/P ratio affected calcium and phosphate utilization and excretion. Both calcium and phosphate became less digestible as the mineral content increased via the meat and bone meal. Phosphate excretion in the feces and urine was decreased by 73% as dietary mineral content was changed from high to low. A low Ca/P ratio increased phosphate absorption, as Table 4.23 shows.

TABLE 4.23 Urinary Excretion of Calcium and Phosphate by Foxes

Ash g/kg	47.00	53.00	70 .00	83.00	97.00	112.00
Ca/P Ratio	1.4	1.9	1.2	1.7	1.2	1.7
Urinary Ca-%	3.4	1.5	0.7	0.4	0.5	0.3
Urinary P-%	35.00	17 .00	35.00	21.00	33 .00	22.00

The fact that a low Ca/P ratio increased urinary phosphate excretion is related to the fact that a low dietary calcium concentration stimulates parathyroid hormone secretion which depresses phosphate reabsorption by the tubules of the kidney (Jongbloed, 1987).

Requirement

The calcium requirement of the growing fox 7 to 37 weeks of age is between 0.5 - 0.6% of dry matter (Harris et al., 1945, 1951b). In the 1951 report it appeared that the optimum level of calcium and phosphorus for growing foxes was 0.6%, each yielding a Ca/P ratio of 1/1. The fox requirement of 0.6% phosphorus was later confirmed by Tauson et al. (1992). The 1951 experiments also indicated that Ca/P ratios in the range of 1.7/1.0 to 1.0/1.0 yielded normal deposition of calcium and phosphate in the fox legs. The authors recommended 0.6 to 1.0% calcium and 0.6 to 0.8% phosphorus with Ca/P ratios of 1/1 to 1.7/1.0.

For optimum lactation performance, Kleckinym (1940) found that silver and blue foxes required about 0.8% calcium and 1.3% phosphorus. His observations further indicated that in the maintenance phase of the ranch year, as little as 0.3% calcium or 0.1 g./100 kilocalories was required.

Resources

An excellent source of calcium and phosphate for fox nutrition is dicalcium phosphate with a Ca/P ratio of 1.1/1.0. Inferior resources include steamed bone meal with a Ca/P ratio of 2.4/1.0 and tricalcium phosphate with a Ca/P ratio of 2.0/1.0. Limestone, that is, calcium carbonate, is a poor source of calcium inasmuch as it contributes to a relatively alkaline urinary pH and thereby is supportive of the urinary calculi, struvite, formation in foxes. Phytic acid, a hexa-phosphoric acid ester of inositol found primarily in the bran portion of cereal grains, is not only a poor resource for phosphate in animal nutrition inasmuch as the acid is bound in an indigestible form but also may combine with free calcium in the intestinal lumen to yield insoluble salts not available to the digestive processes of fur animals.

Nutritional Deficiency

Primary Nutritional Deficiency

A deficiency of calcium or phosphate together with a deficiency of vitamin D can result in the nutritional disease of rickets in young animals. This sub-optimum calcification of the skeleton leads to lameness, recurrent spasms, abnormal skeletal development including enlarged cranial bones, enlarged joints of the front legs which become thickened and crooked, especially at the ankle joint. Forelegs can be as much as 1/4 inch shorter. There were also fractures and edematous swelling of the muzzle. In adults, decalcification of the skeleton causes osteomyelitis, which results in weak and brittle bones (Ott and Coombes, 1941; Harris et al., 1945).

Gorham et al. (1970) described a case of fox nutrition mismanagement. The fox dietary regimen consisted of horsemeat, 75%; horse tripe, lungs and liver, 10%; and cereal 15%. Rickets developed from lack of proper nutrition in three ways: (a) the ratio of calcium to phosphorus was very low; (b) the cereal mixture provided no vitamin D; and (c) there was limited sunlight in the fox shed, minimizing conversion of the skin sterol, 7-dehydrocholesterol, to vitamin D_3. Obviously, the rancher ignored the common sense fur animal nutrition guideline of not to provide more than 25% of the diet with a single fresh/frozen feedstuff.

Secondary Nutritional Deficiency

Even with ample levels of calcium and phosphates in the ranch diet, multiple factors can lead to a secondary deficiency of these critical minerals. Absorption of calcium and phosphates from the intestinal tract depends upon both vitamin D levels and calcium/phosphorus ratios. Vitamin D is required for the absorption of calcium and phosphates from the intestinal tract and essential for bone formation. Experimental studies of Harris et al. (1951) indicated that a control diet that assayed 0.82 IU of vitamin D per gram or 22 IU per 100 kcal of ME was adequate for growing foxes. Rickets can also be induced in fox pups by using two different types of rachitogenic diets involving an excessive mineral imbalance, that is, high or low Ca/P ratios which yield mineral poverty (Ender and Helgebostad, 1949). Dietary programs with high Ca/P ratios provide ample calcium but with insufficient phosphate levels. Compare for instance, the University of Wisconsin Steenbock and Black's rachitogenic diets for rats. With this dietary plan the fox pups are quiet, with no signs of nervousness. These low phosphorus diets are poor in available phosphate, inasmuch as much of the phosphate is bound to phytic acid, which blocks calcium and magnesium salts from absorption in the intestinal tract. In low phosphorus rickets the blood levels of phosphorus are as low as 2-4 mg % versus normal levels of 6-9%. Calcium levels are normal, 9-11 mg %, with a tendency to increase in advanced stages of the disease.

It is of interest to note that Hanson (1935) had warned about the adverse effects of high Ca/P ratios in fox diets, advising that the diets of pups and pregnant and lactating vixens should be supplemented with cod liver oil and edible bone or feeding bone meal. He also warned against the use of limewater, chalk, or limestone which may under certain conditions aggravate conditions rather than alleviating them.

Dietary programs with low Ca/P ratios are poor in calcium but with sufficient or a high content of phosphates. Blood levels of calcium drop to 6-7 mg %, yielding tetany and symptoms of spasmophia, that is, the animals are in a constant state of nervousness and increased reflex irritability. In these low calcium diets the major part of the calcium is withdrawn from absorption by the action of phytic acid yielding the unabsorbable salt, calcium phytate. Even phytin free diets with Ca/P ratios as low as 1/10 with low calcium can produce rickets.

Harris et al. (1951b) noted that Ca/P ratios in the range of 1.7/1.0 to 1.0/1.0 yielded normal deposition of calcium and phosphate in fox legs. Ca/P ratios higher than 1.7/1.0 or lower than 1.0/1.0 can yield abnormal calcium and phosphate deposition in the skeletal structures of foxes and lead to rickets. Studies by Ott and Coombes (1941) and Coombes (1941a, 1941b) indicated that with a vitamin D low basal diet, a Ca/P ratio of 7/1 achieved by the addition of limestone, calcium carbonate, can yield definite rachitic symptoms.

pH of the Intestinal Tract

The absorption of calcium is favored by a low intestinal pH which facilitates keeping the calcium salts in solution. Thus normal gastric hydrochloric acid secretion is necessary for efficient absorption of calcium (Heinz, 1959).

Dietary Binding Factors for Calcium

Mention has already been made of the potential for phytic acid (hexa-phosphoric acid ester of inositol) to bind calcium (Harrison and Mellanby, 1939). Specific fatty acids such as stearic acid yield insoluble calcium salts (soaps) that simply pass through the intestinal tract and are excreted in the feces (Rouvinen and Kiiskinen, 1991).

Toxicity

Ca/P ratios as high as 7/1 bring about rickets in fox pups (Ott and Coombes, 1941; Coombes, 1941a, 1941b).

Chloride

Chloride anion functions in the acid–base balance of the body and is vital for proper osmotic pressure within and without cellular structures wherein chloride and bicarbonate are the major anions of the extracellular fluids. Gastric secretion includes hydrochloric acid and chloride salts.

The nutritional requirements for chloride anion are closely associated with the sodium cation as salt. Thus chloride requirements of the fox will be discussed under the topics of sodium and salt requirements.

Chromium

Chromium has been shown to be an essential nutrient for animals, as it is involved in the functioning of insulin in the regulation of carbohydrate metabolism. Chromium deficient animals have impaired glucose tolerance, that is, a reduced capacity to metabolize glucose. Chromium deficiency symptoms are similar to those of an insulin insufficiency, that is, diabetes.

A chromium deficiency in foxes has not been reported.

Cobalt

There is no scientific data indicating that cobalt as a mineral is an absolute requirement for fox. At the same time, fox require vitamin B_{12}, cobalamine, which contains 4.3% cobalt.

Copper

Copper is a constituent of several metalloenzymes such as cytochrome oxidase, which functions in cellular respiration, and lysyl oxidase, important in connective tissue formation. Copper is also a constituent of tyrosinase, an enzyme involved in the synthesis of melanin, the black pigment of skin, hair, and fur. In terms of the synthesis of blood components, copper is associated with the metalloprotein ceruloplasmin, which functions in iron absorption. Thus, a deficiency of copper may result in an induced iron deficiency. Copper is also necessary for the synthesis of hemoglobin and for red blood cell maturation. Thus a copper deficiency may result in an anemia which is actually an iron deficiency, that is, a secondary iron deficiency directly related to a copper deficiency in the animal's diet. Obviously, copper is a key trace mineral in fox nutrition in terms of its role in hemoglobin formation and normal pigmentation of the fur.

Requirement

There are no experimental reports in the scientific literature on the copper requirements of the fox. Studies at Cornell University with mink on purified diets indicated that 20 ppm copper was satisfactory for growth and fur development (McCarthy et al., 1966).

Fluoride

Fluoride is important in the nutrition of animals inasmuch as it is normally present in bones and teeth. A proper intake is essential for achieving maximum resistance to dental caries.

Physiology

No specific biochemical role for fluoride in animal physiology has been found.

Requirement

No specific experimental data are available on the requirement of the fox for fluoride although a physiological need for fluoride has been demonstrated in other animals.

Resources

Fish are an important source of fluorine, and most meat and cereal products contain significant quantities of fluoride.

Toxicity

Two reports of "apparent" fluoride toxicity in silver fox by Eckerlin et al. (1986, 1988) are of interest. Their field observations were not supportive of fluoride toxicity in foxes at the levels of fluoride present in the commercial fox pellets (National Fur Foods) employed. In fact there is no sound evidence of fluoride toxicity inasmuch as the experimental design did not provide a control group, an experimental group with identical genetics, ranch environment, and ranch management. In their 1986 report, this point is acknowledged in terms of the fact that they felt that evidence of infectious disease or poor management could not be found and thus a causal relationship between fluoride and high kit mortality was suggested.

This report neglects to mention that no other instances of lactation failure had been reported in fox fed the identical formulation. Furthermore, Osweiler et al. (1976) discounts any direct relationship between fluoride toxicity and lactation stating, "Chronic fluorosis does not affect reproduction or milk production directly although the physiological decline in the well being of the affected animals may result in impaired reproduction and reduced milk yield."

The National fox reproduction pellet employed in the 1986 report provided an average of 120 ppm fluoride (range 98-137) and the lactation pellet provided an average of 107 ppm fluoride (range 107-108). An experimental study by Chausow et al. (1992) indicated no agalactia with a fluoride concentration of 254 ppm in National reproduction and lactation pellets over a three year period, even with a level of 354 ppm. Pup survival rate at four weeks was adversely noted only during the third litter. One group of vixens provided 704 ppm fluoride for a single reproduction/lactation period had pups with reduced survival rates but neither litter size nor lactation performance was affected. This level also had no adverse effect on growth or fur density, color, or texture. Neither skeletal nor dental signs of fluoride toxicosis were observed.

Iodide

Iodine is a component of the thyroid hormones thyroxine and triiodothyronine involved in the regulation of cellular metabolism.

Requirement

There are no experimental reports in the scientific literature on the iodide requirements of the fox. Studies at Cornell University with mink on purified diets indicated that 4.1 ppm iodide was satisfactory for the growth and fur development (McCarthy et al., 1966).

Resources

The best fox nutrition resources for iodide would be seafood products, iodized salt, and trace mineral mixtures commonly used in quality commercial fox cereals.

Nutritional Deficiency

There have been no experimental reports in the scientific literature on an iodide deficiency in fox.

Toxicity

No experimental data or field observations have been reported in the scientific literature on iodide toxicity in fox.

Iron

Iron is a key element in biological pigments including hemoglobin, the oxygen carrier in the blood which can also act as an iron storage resource, and myoglobin, an oxygen storage protein in the muscles. It is also involved with the cytochromes and certain enzymes such as catalase and peroxidase. Of special interest to fur nutrition scientists is the fact that iron is an essential element for melanin pigments. Studies by Stout et al. (1968) with radioactive iron (Fe-59) indicated that 70% of the iron in dark mink fur is associated with melanosomes.

Physiology

When fur bearing animals are fed aneamogenic fish, studies by Adair et al. (1962) indicated that mink excreted about 50% more urinary iron and 20% less fecal iron. Thus, a factor in aneamogenic fish provides a significant interference in the retention and utilization of iron.

Requirement

The minimum iron requirement of foxes has not been determined. For most practical fox ranch diets, there should be no problem of an iron deficiency. Quality commercial fox cereals and pellets provide ample levels of iron via ferrous sulfate and chelated iron compounds. The iron nutrition of foxes becomes a problem when aneamogenic fish are provided as part of the ranch fox's dietary regimen as mentioned above. A secondary deficiency of iron can occur wherein ample levels of iron in the fox's diet do not prevent a microcytic, hypochromic anemia.

Resources

Ferrous sulfate, used in human medicine for the treatment of iron deficiency, and iron chelates with amino acids/proteins such as ferrous fumerate or ferrous glycinate and ferric glutamate, a combination marketed as "Hemax," are excellent iron resources. An iron compound of special interest to fox ranch diets including aneamogenic fish is Ferroanemin (Fe-diethyl aminopentacetic acid) which is a complex iron compound inert to the TMAO compounds present in specific aneamogenic fish (Perel'dik and Perel'dik, 1980).

Unfortunately, some inferior iron resources are being employed in commercial fortified cereals for fur animals. These include ferric sulfate, ferrous carbonate, and ferric oxide. Ferric sulfate is used in water purification programs. On addition to water it yields ferric hydroxide, a product that precipitates as insoluble particles. Ferrous carbonate has only 50% of the value of ferrous sulfate for cats (Rogers et al., 1986). Ferric oxide, or iron oxide, has very little bioavailability for animals. In fact, it has been employed for many years in animal nutrition research as a marker in digestibility studies, that is, as a product that is essentially 100% non-biologically available to animals.

Experimental studies have noted that a number of nutrients have the capacity to enhance iron absorption from the intestinal tract, including vitamin B_{12} and the amino acid cysteine. Studies by Taylor et al. (1986) indicate that the value of blood or of meat in non–heme iron absorption is directly related to the iron-chelating effect of cysteine and cysteine containing peptides that are released during animal digestion of meat resources.

Nutritional Deficiency

Primary Nutritional Deficiency

The classic iron deficiency in animals is a microcytic, hypochromic anemia. In the case of foxes, depigmentation of the underfur also occurs (Rimeslatten, 1959). A primary nutritional deficiency of iron (inadequate levels of iron in the animal's diet) in practical fox ranch diets has never been reported.

Secondary Nutritional Deficiency

A secondary nutritional deficiency of iron (ample levels of iron in the dietary regimen but a microcytic hypochromic anemia with depigmentation of the underfur) has been reported in mink fed aneamogenic fish wherein specific dietary factors bind iron in a structure unavailable to the digestive processes of the animal. Commonly used aneamogenic fish include the following:

Coal fish	*(Gadus virens)*	Atlantic Hake	*(Merlucius vulgaris)*
Blue Whiting	*(Gadus potassium)*	Pacific Hake	*(Merlucius merlucius)*
Atlantic Whiting	*(Gadus merlangud)*	Silver Hake	*(Merlucius bilinearis)*
Haddock	*(Gadus aeglefinus)*	Alaskan Pollock	*(Pollachius virens)*

These fish may contain TMAO (Tri-Methyl-Amine-Oxide) or formaldehyde, both of which bind iron in a structure unavailable to the digestive processes of animals (Rouvinen et al., 1997). Studies by Amano and Yamada (1964) indicated that TMAO is broken down by enzymatic action in the fish digestive tract to formaldehyde, a compound known to inhibit iron absorption from the intestinal tract. An increased requirement for iron has been observed in silver and blue foxes fed air-dried cod (Rimeslatten, 1959).

Magnesium

Although about 70% of an animal's body magnesium occurs in the skeletal structure, magnesium is also a key component of the soft tissues. Cardiac and skeletal muscle as well as nervous tissue depend on a proper balance between calcium and magnesium cations for normal function. Magnesium is an essential co-factor for the phosphate-transferring enzymes myokinase, diphosphopyridine nucleotide kinase, and creatine kinase. It is also a co-factor for pyruvic carboxylase, pyruvic oxidase, and the condensing enzyme of the citric acid cycle. Magnesium has a significant role in the diseases of fox inasmuch as it is a key component in fox urinary calculi, struvite (magnesium ammonium phosphate hexahydrate) (Leoschke et al., 1952).

Physiology

The only experimental data available on magnesium physiology in fur animals are those from studies with mink conducted by Leoschke (1960, 1962b). These experiments indicated that about 9-18% of the magnesium intake is excreted via the kidneys. The initial study indicated that magnesium present in bone is not readily available to the digestive processes of the mink.

Requirement

There are no experimental reports in the scientific literature on the magnesium requirements of fox. Studies at Cornell University with mink on purified diets indicated that 0.062% magnesium was satisfactory for growth and fur development (McCarthy et al., 1966).

Resources

Fish are excellent magnesium resources for the fox, and cereal grains provide good resources.

Nutritional Deficiency

Primary Nutritional Deficiency

There are no reports on experimental magnesium deficiency in the fox.

Secondary Nutritional Deficiency

Kangas (1976) emphasized that high levels of calcium and phosphate can decrease the absorption of magnesium from the intestinal tract. In studies only with rats, Forbes (1963) indicated that a magnesium deficiency could be produced only via the addition of both calcium and phosphate to diets already low in magnesium. These additions produced an interference in magnesium absorption leading to high calcium/magnesium ratios in the tissue and led to severe kidney calcification with high phosphate and low magnesium diets.

Toxicity

Magnesium is antagonistic to calcium and phosphate, thus excessive levels of magnesium in fox ranch diets could lead to an inhibition of calcium and phosphate absorption from the intestinal tract (Glem-Hansen, 1978).

Manganese

Manganese functions primarily in the formation of the organic mucopolysaccharide matrix of the bone. Manganese is a co-factor for a galactotransferase enzyme that participates in the formation of the mucopolysaccharide. It is also involved in critical functions of amino acid metabolism.

Requirements

There are no experimental reports in the scientific literature on the manganese requirements of the fox. Studies at Cornell University with mink on purified diets indicated that 61 ppm manganese was satisfactory for growth and fur development (McCarthy et al., 1966).

Molybdenum

Molybdenum is an essential nutrient, being a constituent of the enzyme xanthine oxidase, a biological catalyst involved in the catabolism of nucleic acids and the formation of uric acid. Molybdenum is a trace element of major significance in animal nutrition in terms of its relationship with copper: elevated concentrations of molybdenum can induce a secondary copper deficiency, while, in contrast, low levels of molybdenum can be associated with copper toxicity.

Requirements

At the present time, there are no experimental reports in the scientific literature on the molybdenum requirements of the fox. Studies at Cornell University with mink on purified diets indicated that 1.5 ppm molybdenum was satisfactory for growth and fur development (McCarthy et al., 1966).

Potassium

In contrast to sodium, potassium occurs mainly in the intracellular fluid. The cellular elements of the blood contain about twenty times more potassium than the plasma does; skeletal muscle contains about six times more potassium than sodium. Potassium influences the contractility of smooth, skeletal, and cardiac muscle and profoundly affects the excitability of nerve tissue. Potassium is also important for the fluid and ionic balance of the body.

Requirements

There are no experimental reports in the scientific literature on the potassium requirements of the fox. Studies at Cornell University with mink on purified diets indicated that 0.58% potassium was satisfactory for growth and fur development (McCarthy et al., 1966).

Selenium

Selenium is a constituent of the enzyme glutathione peroxidase which catalyses the breakdown of hydrogen peroxide and other peroxides formed in metabolism. In animal physiology, selenium and vitamin E (alpha tocopherol) are a working pair. Vitamin E, as a natural antioxidant, acts to prevent the development of peroxide structures, while selenium, as a co-factor for glutathione peroxidase , functions in the destruction of peroxides that may occur.

Requirement

There are no experimental reports in the scientific literature on the selenium requirements of fox. An experimental study by Mertin et al. (1991) indicated that supplementing a silver fox diet with a selenium salt resulted in an increase in the length of underfur hairs, most markedly in female foxes. Studies at Cornell University with mink on purified diets indicated that 0.25 ppm selenium was satisfactory for growth and fur development (McCarthy et al., 1966).

Sodium/Salt

Since the nutritional requirements of the chloride anion are closely associated with the sodium cation, the chloride requirements of the fox are being discussed at this point.

Sodium cations and chloride anions are important in the regulation of the acid-base balance of body fluids and determine to a large extent the osmotic pressure of the extracellular fluids. Chloride anions are concentrated by gastric cells in the stomach lining and secreted as hydrochloric acid, important for protein and carbohydrate digestion.

Physiology

In human physiology under normal conditions, 90% of the ingested sodium is excreted in the urine in the form of sodium chloride and sodium phosphates.

Requirement

In the absence of data on the specific salt requirements of foxes, the National Research Council (USA) has recommended that the diet be fortified with 0.5% salt on a dry weight basis (Travis et al., 1982).

Resources

For the most part, the natural content of sodium and chloride in the fox ranch diet feed ingredients should provide ample levels of salt, especially with the recommended supplementation of the diet with 0.5% salt, dry weight basis.

Nutritional Deficiency

No specific experimental observations on a sodium or chloride deficiency in fox have been reported.

Toxicity

For laboratory diagnosis of sodium chloride poisoning, a study by Kacmar et al. (1980) is of interest. Using paleography, the study assessed the physiological concentration of sodium chloride in the livers of foxes on standard ranch diets. The sodium chloride in the livers of polar foxes was 2.5-4.2 g/kg and that of silver foxes was 3.3-3.4 g/kg.

Zinc

Zinc is a key component of multiple metalloenzymes including plasma alkaline phosphates, alcohol dehydrogenase, and thymidine kinase. Thymidine kinase is involved in the ATP (Adenosine TriPhosphate) activation of thymidine, one of the initial steps of DNA polymer construction. A zinc deficiency causes limited cell division which results in dramatic effects on tissues with a rapid growth rate, such as the skin. Dermatitis is a classic sign of zinc deficiency in animals.

Requirement

There are no experimental reports in the scientific literature at the present time on the zinc requirements of the fox. Studies at Cornell University with mink on purified diets indicated that 40 ppm zinc was satisfactory for growth and fur development (McCarthy et al., 1966).

As, Ni, Si, Sn and Va

With animals on purified diets in a sterile, dust-free environment, arsenic, nickel, silicon, tin, and vanadium must be provided to animals for top growth performance and hair or feather development (Frieden, 1972).

5

Disorders Related to Nutrition Mismanagement

Acid/Base Imbalance

A study by Heggset (1998) involved 370 blue fox females for a period of 14 days before and 14 days after whelping. The control diet had a urinary pH range of 5.5 - 5.7 while the experimental diet supplemented with formic acid had a urinary pH of 5.0. The incidence of metritis was significantly lower in fox fed the diet with added formic acid.

Phosphoric Acid Supplementation

A study by Leoschke (1956) indicated that a level of 2.0% phosphoric acid (75% feed grade) dry matter basis provided mink with a urinary pH < 6.0 and was very effective in preventing bladder stone development throughout the ranch year, in addition to providing very significant feed preservation of the mink feed "on the wire." On the basis of this experimental study, National Fur Foods has marketed fox pellets containing 2.0% phosphoric acid (75% feed grade) for a period of more than 20 years as a prophylactic program for struvite urinary calculi. Following Leoschke (1996), the company raised the level of phosphoric acid (75% feed grade) to 2.5% in direct response to fox pellet programs with the 2.0% level of acid yielding bladder stones in fox pups. Very likely these fox pellet programs involved local high alkaline water supply. It is of interest to note that practical mink ranch programs with 2.5% phosphoric acid (75% feed grade), dehydrated basis, provided mink with a urinary pH in the range of 5.1-5.2.

Studies by White et al. (1992) relative to the prevention of chronic struvite urolithiasis in foxes in Nova Scotia involved an even higher level of phosphoric acid in fox ranch diets-2.5% of phosphoric acid (85% food grade) equivalent to 2.8% of phosphoric acid (75% feed grade).

Bladder Stones–Urinary Calculi

History

Urolithasis is relatively common in ranched fur bearers. The condition as seen in mink and fox is similar to LUTD (Lower Urinary Tract Disorder) noted in domestic cats, dogs, and ferrets.

Bladder stones have been responsible for multiple deaths of fox since the very beginning of the mass production of fox as a domesticated animal in North America. As early as 1936, a Virginia fox rancher, R. D. Harmon (1954) experienced a significant loss of fox pups with urinary calculi when unfortunately he made a decision to move from a "homemade" cereal without bone meal or limestone to a commercial fox fortified cereal with 3% steamed bone meal and limestone. With the wise decision to return to his original simple fox cereal without steamed bone meal or limestone, his losses of fox pups with urinary calculi simply ended.

There was no actual requirement of calcium resources in commercial fox cereals at that time, inasmuch as the standard fox diet provided ample levels of green ground bone and fish bone. The fox nutrition scientists working for commerical fox cereal companies simply did not have the foresight to take fox rancher surveys on the actual calcium content of current common fox diets. To put it simply, they did not "Listen To The Fox" —enough said.

Composition

Struvite (magnesium ammonium phosphate hexa–hydrate) uroliths are the most common form of urinary calculi found in foxes. The bladder stones may be almost pure struvite, while other calculi may contain as much as 10-30% cystine in combination with 70-90% struvite (Kechkaylo, 1991).

Etiology

Struvite salts crystalize from the urine in the pH range of 6.0-7.5. The lower the pH of the urine, the greater the proportion of the salt in solution. In vitro studies have shown that the compound is quite soluble in water at pHs below 6.0; however, there is a sharp drop in the solubility of struvite over a pH range of 6.0 to 7.5 (Vermeulen et al., 1951).

Three factors—very specific conditions of pH; urine concentrations of magnesium and ammonium cations, and phosphate anions; and a nucleus, such as cellular debris or bacteria—are required for stone formation in the kidneys or bladder. The potential for calculi is increased with a vitamin A deficiency wherein metaplasia of the pelvic and ureteral mucosa yields large plaques of desquamated epithelium which may act as niduses for the development of stones. Bassett et al. (1946) noted that fox fed low levels of vitamin A had a high incidence of urinary calculi.

Sub-optimum levels of phosphoric acid employed for the prevention of urinary calculi in fox may actually predispose the fox to bladder stone formation. The lower levels of phosphoric acid in the fox dietary regimen enhance the urinary levels of phosphate anions required for struvite (magnesium ammonium phosphate hexa-hydrate) formation, but yield a urinary pH too high to solubilize the struvite components. While the level of 2.0% phosphoric acid (75% feed grade) in National Fox pellets has prevented the formation of struvite calculi over the years, the employment of 1% phosphoric acid (75% feed grade) in a commercial fox pellet led to significant losses of male foxes with bladder stones (Mosshammer, 1985).

Urinary Calculi—Prophylactic Treatment–Urinary Acidifiers

Phosphoric Acid

With the success of phosphoric acid for the prophylactic treatment of struvite bladder stones in mink (Leoschke, 1956), National Fur Foods Company started employing 2.0% phosphoric acid

(75% feed grade) in its fox pellet formulations. However, after significant losses of male breeder foxes with urinary calculi on this formulation, the level of the acid was increased to 2.5% in 1994 for the National Fox Reproduction pellets. This change apparently resolved the problem. This decision to increase the level of phosphoric acid was supported by the findings of White, et al. (1992) relative to the employment of 2.5% phosphoric acid (85% food grade) on the prevention of chronic struvite urolithiasis. For the National Fur Foods Company Super-Gro fox formulations, the level of 2.0% phosphoric acid (75% feed grade) was maintained. In 1989, a decision was made to remove the phosphoric acid supplementation of the National Fox Lactation pellets as the direct result of rancher complaints of low palatability for the lactating vixens (Kidder, 1988). This change in formulation resolved the problem of low palatability. Since the major problem of bladder stones was in male breeder foxes and male pups, the change had no effect on the prophylactic treatment for urinary calculi in these animals.

Ammonium Chloride/ DL-Methionine

Westlake (1989) has recommended a combination of 24 ounces of ammonium chloride and 6 ounces of DL-methionine per 800 females/day for the prevention of struvite urinary calculi. The DL-methionine contributes to urinary acidity via its metabolism to hydrogen sulfate anions. This dosage is 6X those recommended for mink, since foxes weigh about 6X more than mink.

Ammonium Chloride or Hydroxy-Methionine

White et al.(1992) have conducted studies on the employment of a fox pellet supplemented with either 1.0% ammonium chloride or 1.7% Alimet (a methionine hydroxy analog) for the prevention of chronic urolithiasis.

6

Modern Fox Nutrition Management

Fortified Cereals

Relatively few experimental reports have been published on the nutritional value of various cereal grains or cereal by-products for the fox. Thus, I offer the following brief commentary.

Oats–Dehulled/Naked

Studies by Nenonen et. al. (2002) indicate that the barley portion of a blue fox ranch diet can be replaced with either dehulled oats (oat groats) or naked oats, with equal performance of the fox. These products are superior to barley in terms of possessing a higher fat content and smaller quantities of soluble beta-glucanes, which are not readily digested by fur animals.

Barley

Experimental studies conducted by Valaja et al. (2003) indicated that a blue fox diet containing 28% barley (1/2 raw with storage for a week with supplemental beta-glucanase enzyme) provided growth and fur quality equal to a control diet containing 11% cooked barley. Fecal quality was satisfactory.

Corn Germ Meal

In an unauthorized field experiment (1987), National Fur Foods Company marketed a Super-Gro fox pellet containing a 15% level of corn germ. The net result was that nine fox ranchers feeding that pellet observed significant problems with diarrhea in their foxes. Curiously, in one instance, no problems of diarrhea were noted when the pellet was fed as a wet mash to fox in outside pens, but diarrhea problems were noted when these same fox were provided the experimental pellet as a dry pellet in inside pens. Studies conducted by Anderson (1988) indicated a Gross Energy (GE) digestibility of only 61% for corn germ meal, the lowest GE digestibility value of all corn products tested including raw corn, extruded corn, and corn gluten meal.

Soybean Products

For modern fox nutrition, soybean oil meal products have the advantage of a relatively low cost per protein unit. However, the protein has a relatively low level of methionine, a key sulfur amino acid for fur production. Another problem is that raw soybean oil meal contains an anti-trypsin factor which inhibits the digestion of proteins in the intestinal tract. Fortunately, this factor is unstable to heat treatment. Suitable heating processes denature the anti-trypsin factor in soybean oil meal thus enhancing the digestion of protein and, what is especially important, the breakdown (hydrolysis) of protein peptide bonds to yield free sulfur amino acids, cysteine and methionine. However, all factors considered, there is some question of whether the current heating practices provide the highest quality of soy protein for modern fur nutrition.

The process of heating raw soybean oil meal after pressing to remove the oil content yields a product known as soya cake. In North America, the most common soybean oil meal product is a 48% protein resource prepared by solvent–extracting the soybean oil meal to remove the oils and then heating to denature the anti-trypsin factor.

Raw Soybean Meals

The employment of raw soybean meal in silver fox diets may yield poor performance of the animals. Studies by Rimeslatten as reported by Hoie (1953) indicated that the use of raw soya meal at 15-20% of the energy level resulted in lower weight gains of silver fox cubs. With larger quantities of raw soybean meal in the fox diet, as high as 25-40 grams/fox/day, the palatability of the diet was reduced, leading to even lower weight gains.

With raw soya meal in the diet of lactating silver foxes at a level of 20 grams/fox/day, 35% of the cubs died at 3-4 weeks.

Soya Cake

Rimeslatten (1974) reported that silver foxes on a diet containing 40% of total protein from soybean cake had slightly lower final body weights but an increase in total pelt value. In another study with blue foxes with soya cake as 21% of total protein for the cubs, weight gains were equal to those of the control animals with only an insignificant decrease in pelt prices.

In the Rimeslatten studies, the droppings of the fox cubs on both raw soybean meal and soya cake were somewhat softer. This observation is of interest inasmuch as Leoschke (1983) noted reports from fox farmers providing their animals with a new unauthorized "experimental" National Fur Foods fox pellets with approximately 25% soybean oil meal (solvent extracted and heated) that the cubs exhibited diarrhea. The problem was immediately resolved with new shipments of regular commercial fox pellets with a lower level of processed soybean oil meal.

Soybean Oil Meal-48

Perel'dik et al. (1972) noted that 40% of the protein during the summer and fall months and 20% of the protein during the reproduction/lactation phase of the fox ranch year could be provided as soybean oil meal protein without adverse effects. In other studies, it was noted that with 50% of animal protein replaced with soybean protein, there was a decrease in protein digestibility of 4-5%.

A study by Makela and Kiiskinen (1975) with blue foxes indicated that with 35% of digestible protein replaced with soybean oil meal protein there was a 9% lower growth from July to pelting but superior fur quality relative to the animals within the control group.

Dehydrated Protein Resources

Bacterial Proteins

Experimental studies with blue foxes by Skrede and Ahlstrom (2002, 2004) indicate that a bacteria protein meal (BPM) produced by continuous bacterial fermentation with natural gas as a carbon and energy source may prove to be a unique resource of protein for modern fox nutrition in pellet form. The research data indicated that no significant differences in digestibility of protein, fat, or carbohydrates were noted with increasing levels of BPM. It is of interest to note a tendency toward enhanced growth rate with increasing levels of the product—obtained without any increase in feed volume—indicating a marginal improvement of feed conversion. No significant differences in fur quality were noted among the treatment groups.

Meat-And-Bonemeal

Studies by Skrede and Ahlstrom (2001) indicated that meat-and-bone meal from carcasses of farm animals (50% crude protein with a mink digestibility coefficient for protein of 60%) could replace fish meal at a level of 22.5% of total digestible protein during the critical reproduction/lactation phase of the fox ranch year and still maintain top performance of the animals. However, at the level of 45% of total digestible protein, pup body weight at weaning was significantly sub-optimum.

7

Practical Fox Ranch Diets–20th Century

Recommendations of the National Research Council (USA) (Travis et al., 1982) are provided in Table 7.1.

TABLE 7.1 Suggested Ranges of Composition of Practical Diets for Foxes

Ingredients	**Percent**
Fortified Cereal*	25-50
Liver	0-10**
Quality Proteins***	5-30
Cooked eggs, whole poultry, whole fish,	
horsemeat, rabbits, nutria, blood, etc.	
Beef By-Products	10-20
Tripe, lungs, lips, udders, spleen, etc.	
Poultry By-Products	0-50
Heads, entrails, feet	
Fish Scrap	0-50
Fishmeal	5-15
Fat Supplementation	0-10****
Rendered animal fat or vegetable oils	

Proximate Analysis***	**Percent**
Protein	20-30
Fat	15-30
Carbohydrates	25-60
Ash	5-15

* May consist of single cooked grains such as oat groats or wheat in combination with vitamin and trace mineral supplementation or commercially prepared fortified cereal mixtures.

** Reproduction/lactation diets (March-May) often contain 5-10% beef liver, although necessity for this has not universally been accepted.

*** Level of quality protein feedstuffs is often increased during the critical fur development and reproduction/lactation phases, a practice consistent with the higher protein requirements of foxes during these periods.

**** A level of fat supplementation that provides proper protein/energy balance for each phase of the ranch year.

***** A proximate analysis consistent with the optimum balance for each phase of the life cycle.

8

Modern Fox Nutrition—21st Century–Pellets

National Fur Foods Company was marketing inferior fox pellet products when I became associated with the firm in 1955 as Director of Fur Animal Nutrition Research. However, I was not asked to review the fox pellet formulations until almost a quarter century later in 1979, and then only after fox ranchers complained about second rate performance of their animals on the National Fox Pellet Program in terms of size, and especially, sub-optimum lactation. I was able to upgrade the National fox formulations to produce higher energy levels and enhanced protein nutrition in terms of both protein quantity and protein quality. The revisions were based on (1) experimental fox data provided by Hans Rimeslatten at the first International Congress in Fur Animal Production held in Helsinki, Finland in April, 1976, and (2) knowledge I had obtained in the previous decade on the biological value of dehydrated protein feedstuffs for the mink in experimental studies at the National Research Ranch facilities.

Growing/Furring Phase: The original National fox formulation for this period provided only 28% protein and 10% fat. The new Super-Gro fox pellet provided 36% protein and 17% fat, as well as dehydrated protein feedstuffs with higher biological value for fur animals. The immediate result was a fox pellet supportive of enhanced body weights at pelting and superior fur quality.

Lactation Phase: The original National fox formulation for this phase of the ranch year provided 27% protein and 7% fat, while the new pellet provided 31% protein (upgraded in terms of both quantity and quality via protein resources of higher biological value) and 10% fat . Within a year my efforts to upgrade the National fox lactation pellet were rewarded by the remarks of a fox farmer at the International Mink Show. He simply stated three little words—not "I love you," but words much more important to a fur animal nutrition scientist, "Wonderful milk bags!"

Modern Fox Pellet Formulations

The nutrition of the fox is much simpler than that of the mink. Fox were being raised on pellet formulations 75 years ago with superior fur development in terms of sharp color while mink pellets supportive of top performance of the mink were not marketed until almost 50 years later. The greater challenge to fur animal nutrition researchers to produce a top performance mink pellet required a longer period of experimental studies.

The composition of a number of modern fox pellets is provided in Table 8.1. Quality commercial fox pellets support superior color in pelts marketed because all the fats provided in the formulations are stabilized by Vitamin E and powerful antioxidants. These fat resources arriving on the surface of the fox fur are resistant to a process termed "oxidative rancidity" that yields "off color" pelts. This effect has been known for more than 75 years: Smith (1932), extolling the nutritional value of Purina Fox Chow Checkers, offers the following commentary:"Outstanding feature was the color of the pelts, which was far superior to that of the other pelts on the same sale. The lustrous black color of the pelts and the absence of brown rusty color were the interesting features of the pelts."

Experimental data from the modern era of fox nutrition management supports the basic concept of using a commercial pellet for superior fur development, as Table 8.3 shows.

TABLE 8.2. Modern Fox Pellet Formulations

Nutrients	Breeding		Growing-Furring	
	Danish*	NovaScotia**	Norway***	Denmark
Protein Resources				
Fishmeal – low ash	36.6	20.0	3.0	15.5
Corn Gluten – 60%	7.0	8.4	10.0	
Meat & Bone Meal	1.2	6.0	3.0	10.4
Poultry By-Product Meal			3.0	
Bloodmeal		1.5		
Toasted Soybean Meal	3.0	1.5	7.8	11.4
Potato Protein Conc.	5.0	5.0		
Dried Whey			4.0	
Methionine	0.22			
Lysine	0.03			
Carbohydrate Resources				
Extruded Whea	20.4	17.4	28.5	43.6
Wheat Middlings	4.1	4.5	25.0	
Extruded Corn		5.0		
Potato Powder		5.0		
Glucose	5.0			
Molasses			1.0	
Fat Resources				
Beef Fat (Tallow)			13.7	
Swine Fat (Lard)	5.5			
Soybean Oil	3.5		14.8	
Vegetable Fats		13.4		
Fortification				
Minerals/Vitamins	3.5	2.5	0.5	0.3
Salt	0.25	0.25		
DiCalcium Phosphate		1.0		
Calcium Carbonate		1.0		
Anti-Oxidants			0.5	
Fibre				
Beet Pulp	4.8	3.5		4.0
Green Meal		4.0		

Proximate Analyses			
Protein	42	33	29
Fat	13	17	21
Ash	8	8	6
Fibre	3	3	4
Energy Calories/kilogram	3270	3290	

* KFK **White et al. (1992) ***Skrede and Ahlstrom (2002, 2004)

TABLE 8.3. Growth/Fur Quality Performance—Ranch Mix vs. Pellet Program*

Dietary Program	**Fur Quality Blue Foxes**				
	A	**B**	**C**	**D**	**AVG**
Males					
Ranch Mix	7	16	13	0	2.7
Pellets	6	13	16	1	2.4
Females					
Ranch Mix	3	18	8	3	2.4
Pellets	5	21	6	0	2.9

Dietary Program	**Fur Quality Silver Foxes**						
	A+	A	B+	B	C+	C	AVG
	6	5	4	3	2	1	
Males							
Ranch Mix	1	15	6	37	2	11	3.2
Pellets	4	30	5	29	0	4	4.0
Females							
Ranch Mix	0	5	6	26	3	15	2.7
Pellets	1	16	2	31	2	3	3.5

*Leoschke and Chausow (1988)

References

Aarstrand, K. and Skrede, A., (1993). Effect of different storage methods on quantities and chemical composition of manure from fur bearing animals. Norsk Landbruks Forskning 6: 339-358. ISSN 0801-5333 (Scientifur 17(1):34

Abramov, M.D. and Poveckij, I.G. (1955). Standards and rations for adult blue foxes. Karakul. Zver. 8(1): 26-28.

Ackerman, A. (1985). East State Veterinarian Clinic, Fort Wayne, Indiana. Personal communication.

Adair, J., Stout, F.M., and Oldfield, J.E. (1962). Experiments in mink nutrition - 1961 Progress Report. Oregon State University.

Afanas'ev. V. A. (1957). Sealmeat, a good protein feed for young fur animals. Karakul. Zver., 10, No. 4: 28-30.

Ahlstrom, O. (1992). Different dietary fat:carbohydrate ratios for blue fox in the reproduction period. Effects on reproduction, kit growth, milk composition and blood parameters. Norwegian J. of Ag. Sci. Suppl. #9 (1992). Progress in Fur Animal Science. Proceedings of the 5th International Congr.In Fur Animal Production pp. 242-248.

Ahlstrom, O. and Skrede, A. (1995a). Comparative nutrient digestibility in blue foxes (Alopex lag opus) and mink (Mustela vison) fed diets with diverging fat:carbohydrate ratios. Acta Agr. Scand. 45(1): 74-80.

Ahlstrom, O. and Skrede, A. (1995b). Fish oil as energy source for blue foxes

(Alopex lagopus) and mink (Mustela vison) in the growing-furring period. J. Anim. Physiol. A. Anim. Nutr. 74:146-156.

Ahlstrom, O. and Skrede, A. (1995c). Feed with divergent fat:carbohydrate ratios for blue foxes (Alopex lagopus) and mink (Mustela vison) in the growing period. Norwegian J. Agr. Sci. 9: 115-126.

Ahlstrom, O. and Skrede, A. (1995d). Liverlipids and paroxismal beta-oxidation in blue foxes and mink. Scientifur 19(1): 66.

Ahlstrom,O. and Skrede, A. (1996). Marine fat versus carbohydrates as energy sources in diets for blue foxes in the reproduction period. Animal Production Review, Polish Society of Animal Production. Applied Science Reports 28 Progress in fur animal science. Nutrition,Patalogy and Disease. Proceedings from the 6th Inter. Sci. Congr. In Fur Animal Production. August 21-23, 1996, Warsaw, Poland. Ed. A. Frindt & M. Brzoxowski.

Ahlstrom, O. and Skrede, A. (1997). Energy supply for blue fox and mink in the reproduction period. Scientifur 21(4): 300.

Ahlstrom, O., Skrede, A. and Tangen, S.F. (1998). Comparative nutrient digestibility in blue fox (Vulpes vulpes) and mink (Mustela vison). Proceedings NJF-Seminar no. 295. 6 pp.

Ahlstrom, O. and Wamberg, S. (1997). Measurement of daily milk intake of suckling fox cubs. Scientifur 21(4): 304.

Ahlstrom, O., Wamberg, S., Sanson, G. and Tauson, A-H (2000). Measurement of milk production in blue fox dams with different litter size using an isotope dilution technique. Scientifur 24(4) - IV-A - Nutrition. Proceedings of the 7th Inter. Sci. Congr. In Fur Animal Production, ed. B.D. Murphy and O. Lohi: 60-62.

Allison, J.B. (1949). Biological evaluation of protein. Adv. Prot. Chem. 5: 194-196.

Amano, K. and Yamada, K. (1964). A biological formation of formaldehyde in the muscle tissue of Gadoid fish. Bull. Japan Soc. Sci. Fish 30:430-432.

Anderson, D.M. (1988). Digestible nutrients in corn and corn by-product feeds for silver foxes. Biology, Pathology and Genetics of Fur Bearing Animals, ed. B. D. Murphy and D. B. Hunter. Proceedings of the 4th Inter. Congr. In Fur Animal Production, Rexdale, Ontario, Canada. August 21-24, 1988: 382-390.

Anon (1929). Silver Fox Milk (Colostrum). Otakar Laxa Ann. Fals 22: 598-600.

Anon (1930). Silver Fox Milk (5 weeks after parturition. Otaka Laxa Ann. Fals 23: 404.

Anon (1954). Food norm for mink kits. Vara Pelsdjur No.2, Feb 1: 23-28.

Anon (1977). Fiskeens, iagens indflydelse pa passagegetiden hos mink. Stenc.Bilag Til Staten, Husdybrugs-Forsogs-Forsogs Arsmode.

Anon (1993). Effects of different storage methods on quantities and chemical composition of manure from fur-bearing animals. Scientifur 17(1): 34.

Anon (1996). Animal Production Review, Polish Society of Animal Production. Applied Science Reports. 28 Progress In Fur Animal Science. Nutrition, Patology and Disease. Proc. 6th Inter. Sci. Congr. In Fur Animal Production.

Anon (1997). Fur animal production and early research in the Nordic countries. Scientifur 21(4): 291.

Arrington, L. R. and Davis, G.K.(1953).Molybdenum toxicity in the rabbit. J. Nutr. 51: 295-304.

Avery, B. (1947). Yearly Fox Feeding Schedule, National Fur News 19(1): 14.

Barkoll, A. (1987). Indiana veterinarian. Personal Communication.

Basset, C. F. (1951). U.S. Bureau Anim. Exp. Sta., Saratoga Springs, New York. Unpublished data.

Bassett, C. F., Harris, L.E., Smith, S.E. and Yeoman, E.D. (1946a). Urinary calculi associated with vitamin A deficiency in fox. The Cornell Veterinarian 36(1): 5-16.

Bassett, C.F., Harris, L.E. and Wilke, C.F. (1946b). A comparison of carotene and vitamin A utilization by the fox. The Cornell Veterinarian 36(1): 16-24.

Bassett, C.F., Loosli, J.K.and Wilke, C. F. (1948). The vitamin A requirement of growing foxes and minks as influenced by ascorbic acid and potatoes. J. Nutr. 35: 629-638.

Beerbard, R. and Smith, S.E. (1941). Digestion and absorption by foxes and mink. American Fur Breeder XIV (2):22.

Bell, J.M. and Thompson, C. (1951). Freshwater fish as an ingredient of mink rations. Bull. No. 92, Fisheries Res. Board of Canada.

Berg, J., Juokslahati, T. and Makela, J. (1982). Low protein feed for fox. NJF Meeting, Alesund, Norway.

Bernard, R., Smith, S.E. and Maynard, L. A. (1942). The digestion of cereals by mink and foxes with special reference to starch and crude fiber. Cornell Vet. 32(1): 29-36.

Bleavins, M. R. and Aulerich, R. J. (1981). Feed consumption and food passage time in mink (Mustela vison) and European ferrets (Mustela Putorius furo). Laboratory Animal Science 31(3): 268-269.

Borgstrom, G. (1961). Fish As Food. Academic Press, New York, N.Y.

Brody, S. (1935). The relationship between feeding standards and basal metabolism. Report of Conference on Energy Metabolism held at State College, Penn. June l935. p. 12.

Brody, S. (1945). Bioenergetics and Growth. Reinhold Publ. Co. N.Y.

Brothers, A. W. (1985). HiLite Fur Farms Lt., Rockwood, Ontario, Canada. Personal communication.

Brzoowski M. and Zakrzewska-czarnogorska, E. (2004). Influence of using enzymatic preparations: alpha-amylase, beta-glucanase and xylanase on nutrient digestibility in polar foxes (Alopex Lagopus) Scientifur 28(3): VIII Inter. Congr. In Fur Animal Production. DeRuwenberg 's Hertogenbosch, The Netherlands - 15-18, September 2004: 100-1002.

Buddington, R.K., Chen, J.W. and Diamond, J.M. (1991). Dietary regulation of intestinal brush border sugar and amino acid transport in carnivores. J. Physiology 261 (4): Part 2: R793-R801.

Ceh, L., Helgebostad, A. and Ender, F. (1964). Thiaminase in cepelin (Mallotus villous) an Artic fish of the Salonidae family. Interzeitschrift Fur Vitamin Forshung 32(2): 189-196.

Chaddock, T. T. (1939). Report on grey and red fox stomach examinations. Wisc. Conservation Bull. 4(9): 53-54.

Charlet-Lery, G., Fiszlewicz, M., Morel, M.T. and Richard, J. R. (1981). Rate of passage of feed through the digestive tract of mink according to the type of diet (wet form or pellets). Ann. Zootech. 30(3): 347-360.

Chausow, D. G. (1989). National Research Ranch. Unpublished data.

Chausow, D.G., Clay, A.B. and Suttie, J. W. (1992).Effects of dietary fluoride on growth, reproductive performance, and tissue fluoride levels of the fox (Alopex lagopus).Norwegian J. of Agr. Sci. Supple. No. 9. Progress in Fur Animal Science. Proceedings from the 5th Inter. Sci. Congr. In Fur Animal Production.

Chesemore, D. L. (1968). Notes on the food habits of Artic fox in Northern Alaska. Canadian J. of Zoology 46(6): 1127-1130.

Colin, G. (1954). Traite de Physiologic Comparee des Animaux Domestique Tome Premier. Paris, J-B Bailliere: 408-410.

Coombes, A., (1940). Feeding fish to fur bearing animals. American National Fur and Market J. 19(3): 5-6, 24.

Coombes, A. I. (1941a). Nutrition and the proper feeding of foxes and mink.I. The American National Fur and Market J. 20(1): 13-14, 20-21.

Coombes, A. I., Ott, G. L. and Wisnicky, W. (1940). Vitamin A studies with foxes. North Amer. Vet. 21: 601-606.

Cutter, W. L. (1958). Food habits of the swift fox in Northern Texas. J. Mammalogy 39(4): 527-532.

Dahlman, T. (2005). Protein and amino acids in the nutrition of the growing-furring blue fox. Academic Dissertation by Tuula Dahlman. ISBN 952-10-0860-1 (NID), ISBN 952-10-0861-X(PDF), ISSN 1236-9837.

Dahlman, T. and Blomstedt. (2000). Effect of feed protein level on fur and skin of the blue fox. Scientifur 24(4): IV-A - Nutrition: 12-16. Proceed. Of the VIIth Inter. Sci. Congr. In Fur Animal Production.

Dahlman, T., Kiiskinen, T., Makela, J., Niemela, P., Syrjalaqvist, L., Valaja, J. and Jalava, T. (2002a). Digestibility and nitrogen utilization of diets containing protein at different levels and supplementation with DL-methionine, L-methionine and L-lysine in blue fox (Alopex lag opus). Animal Feed Sci. Technol. 98 (3-4): 221-237.

Dahlman, T., Valaja, J., Niemela,P. and Jalava, T. (2002b). Influence of protein level and supplementary L-methionine and lysine on growth performance of blue fox (Alopex lag opus). Acta Agr. Scand. Sec. A. Animal Science #4 52(4): 174-182.

Dahlman, T. and Valaja, J. (2003). Growth and skin quality of blue foxes raised on different protein and amino acid levels. NJF-seminar #354, Lillehammer, Norway.

Dahlman, T., Valaja, J., Jalava, T. and Skrede, A. (2003). Growth and fur characteristics of blue foxes (Alopex lagopus) fed diets with different protein levels and with and without DL-methionine supplementation in the growing-furring period. Can. J. Anim. Sci. 83:239-245.

Dahlman, T., Valaja, J., Niemela, P. and Jalava, T. (2002). Influence of protein level and supplementary L-methionine and lysine on growth performance of blue fox (alopex lag opus). Acta Agr. Scand. Sec. A. Animal Science #4 52(4) 2002: 174-182.

Dahlman, T., Valaja, J., Venalainen, E., Jalava, T. and Polonen, I. (2004). Optimum dietary amino acid pattern ad limiting order of some essential amino acids for growing-furring blue foxes. Animal Science 78:77-96.

Dekker, D. (1983). Denning and foraging habits of red foxes (Vulpes vulpes) and their interaction with coyotes (Canis latrans) in central Alberta, 1972-1981. Can Field Nat. 97(3): 303-306.

De Ritter, E. (1976) Stability characteristics of vitamins in processed foods. Food Technology 30: 48-52.

Deutsch, H.F. and Hasler, A. D. (1943). Distribution of a vitamin B-1 destructive enzyme in fish. Proc. Soc. Exp. Biol. Med. 51: 119-121.

Deutsch, H.F. and Hasler, A.D. (1945). Distribution of vitamin B-1 destructive enzyme in fish. Proc. Soc. Exp. Biol. Med. 53: 63-65.

Dille, L.L., Bakken, M., Eldoy, O.A., Johannessen, K-R., Kassin, S., Moe, R.O., Westerjo, S.G. and Ardal, O.D. (1998). Vannforbruk hos blare I lopet av vinteren. NJF-Seminar 295, Bergen, Norway.

Dixon, J. (1925). Food predilection of predatory and fur bearing mammals. J. Mammalogy 6(1): 34-47.

Dodds, D. G. (1955). Food habits of the Newfoundland red fox. J. Mammalogy 36(2): 291.

Eadie, W. R. (1943). Food of the red fox in Southern New Hampshire. J. Wildlife Management 7(1): 74-77.

Eckerlin, R. H., Krook, G. A., Maylin and Carmichael, D. T. (1986). Toxic effects of food-borne fluoride in silver foxes. Cornell Vet. 76: 395-402.

Eckerlin, R. H., Maylin, G.A., Krook, L. and Carmichael, D. T. (1988). Ameliorative effects of reduced food-borne fluoride on reproduction in silver foxes. Cornell Vet. 78: 385-391.

Elnif, J. and Enggaard-Hansen, N. (1988). Production of digestive enzymes in mink kits. Biology, Pathology and Genetics of Fur Bearing Animals. 4th Inter. Congr. In Fur Animal Pro. August 21-24, l988. Rexdale, Ontario, Canada. 320-326.

Elnif, J. and Hansen, N. F. (2005). Metabolism of D- and L-methionine in mink (Mustela vison). Proc. From NJF-Seminar Nr. 377. 5 pp.

Ender, F. and Helgebostad, A. (1939). Experimental beri-beri in the silver fox. Skand Vet Tidskr. 29: 1231.

Ender, F. and Helgebostad, A. (1943). Undersokelser over beriberi hos rev under dreskjtighet og laktasjon samt forsk pa a bestemme revens vitamin B-1-behov. Nordisk Vet. Tidskr 1: 3-50.

Ender, F. and Helgebostad, A. (1951a). Further investigations on the influence of the food on the quality of the silver fox fur III. The etiological factors in hyperkeratosis of the skin and dandruff in the fur of silver foxes. Saertrykk av Norsk Pelsdyrblad nr. 6 : 3-12.

Ender, F., and Helgebostad, A. (1951b). Continued investigation of the effect of feeding on the fur quality of the silver fox. III. Norsk Pelsdyrblad 25: 193-202.

Ender, F. and Helgebostad, A. (1953). Yellow fat disease in fur bearing animals. Proc. XVth International Vet. Congr. Stockholm, August 9-15th, IV:141.

Ender, F. and Helgebostad, A. (1958). Further studies on the influence of various food constituents on the quality of the fur in foxes and mink. Saertrykk av Norsk Pelsdryblad nr. 1.

Ender, F. and Helgebostad, A. (1959). Experimental biotin deficiency in mink and foxes. Nord. Vet. Med. 11:141-161.

Ender, F., Helgebostad, A. and Bohler, N. (1949). Studies on experimental rickets and tetany in foxes. Nord. Vet.-Med. 1: 827-894.

Enggaad-Hanson (1991). Dietary effects of omega-3 polyunsaturated fatty acids on body fat composition and health status of farm raised blue and silver foxes. Acta Agr. Scand. 41(4): 401-414.

Enggaard-Hansen, N., Finne, L., Skrede, A. and Tauson, A-H. (1991). Energiforsymingen hos mink or raev - NJF Utredning/ Rapport No. 63, DSR (publisher), Agr. Univ. Copenhagen, Denmark, 59 pp.

Errington, P. L. (1935). Food habits of mid-west foxes - Wisc. and Iowa. J. Mammalogy 16(3): 192-200.

Evans, C. A., Carlson, W. E. and Green, R.G.. (1942). The pathology of Chastek paralysis in foxes and counterpart of Wernicke's hemorrhagic polioencephalitis of man. Am. J. Pathology 18(1): 79-92.

Ewos Fox Pellets (1980). Ewos A/S, Stagehojvej 27, 8600 Silkeborg,Sweden

Farrell, D. G. and Wood, A. J. (1968). The nutrition of the female mink (Mustela vison). III. the water requirement for maintenance. can. j. zool. 46(1): 53-56.

Faulkner, W. L. and Anderson, D.M. (1991). The effects of fiber supplementation on diet digestibility of silver foxes. Can. J. Anim. Sci. 71(3): 943-947.

Faulkner, W. L., Egan, L.A. and Anderson, D. M. (1992). Apparent and true digestibility of dry matter, crude protein and amino acids in diets for mature silver foxes. Norwegian J. of Ag. Sci. Supplement #9, 1992. Progress in Fur Animal Science, Proc. 5th Inter. Sci. Congr. In Fur Animal Production, Oslo, Norway: 268-274.

Fay, F. H. and Stephenson, R. O. (1989). Annual, seasonal and habitat related variations in feeding habits of the artic fox (Alopex lag opus) on St. Lawrence Island, Bering Sea. Can. J. Zool. 67: 1986-1994.

Ferns, L.E. and Clark, M.H. (1988). Congestive cardiomyopathy in juvenile ranch foxes (Vulpes vulpes) - report of an outbreak. Biology, Pathology nd Genetics of Fur Bearing Animals. 4th Inter. Sci. Congr. In Fur Anim. Production. 117-182.

Firstov, A. A. and Haritonov, P. A. (1957). Feeding silver foxes during pregnancy. Karakul. Zver. 10, No. 2: 27-30.

Firsov, A.A., Vahameev, K. A. and Sidorov, V. M. (1950). The protein level for young pelting foxes. Karakul. Zver. 3:50-52.

Fleming, P. (1999). Nova Scotia mink rancher - personal communication.

Fog, A. (1974). A vitamin. Bilag til kursus om vitamin-og mineralstoffer i minkens foder, deres indflydelse pa reproduktion, vaekst og pelsudvikling. Tune Landboskloe, Denmark.

Forbes, R.M. (1963). Mineral utilization in the rat. Effects of varying dietary ratios of calcium, magnesium and phosphorus. J. Nutr. 80: 321-326.

Forbes, T. O. A. and Lance, A.N. (1976). The contents of fox scats from Western Irish Blanket Bog. J. Zool. (London). 179:224-226.

Fors, F.M., Valatonen, M., Tyopopponen, J. and Polonen, I. (1990). Foderts energihalt och dess inverkan pa blarevenms valpresultat. Nordisk Jordbruksforskers forening. Seminar 185, Tastrup, Denmark. 8 pp.

Frank, L. G. (1979). Selective predation and seasonal variation in the diet of the fox (Vulpes vulpes) in N.E. Scotland. J. Zoology (London). 192(4):561.

Frieden, E. (1972). The chemical elements of life. Sci. Amer. 227: 52-60.

Glover, F. A. (1949). Fox foods on West Virginia wild turkey range. J. Mammalogy 30 (1): 78-79.

Glem-Hansen, N. (1978). Erfarungen mit dem soyaprotein produkt "Nurupan". Die Mohle + Misch Futterchnik 15:572.

Gnaedinger, R.H. (1965). Thiaminase activity in fish. An improved assay method. U. S. F. W. S. B. B. C. F. Fish Ind. Res. 2(4):55.

Gorham, J. R., Peckham, J. C. and Alexander, J. (1970). Rickets and osteodystrophia fibrosa in foxes fed a high horsemeat ration. American Veterinary Medical Association - Journal 156(9): 1331-1333.

Goszynski, J. (1974). Studies on the food of Polish foxes. Acta Theriologica 19: 1-18.

Green, R. G. (1938). Chastek paralysis. Amer. Fur Breeder XI(1):4-8.

Green, R.G. and Evans, C. A. (1940). A deficiency disease of foxes. Science 92:154-155.

Green, R. G., Carlson, W. E. and Evans, A. (1941). A deficiency disease of foxes produced by feeding fish. B-1 avitaminosis analogous to Wernicke's disease of man, J. Nutr. 21: 243-256.

Green, R.G., Carlson, W. E. and Evans, C.A. (1942a). The inactivation of vitamin B in diets containing whole fish. J. Nutr. 23(2): 165-174.

Green, R. G., Evans, C.A., Carlson, W. E. and Swale, F.S. (1942b). Chastek paralysis in foxes. B avitaminosis induced by feeding fish. J. of the Amer. Vet. Med. Assoc. 100: 394-402.

Gregory, J.F. and Kirk, J.R.(1981).The bioavailability of vitamin B-6 in foods. Nutr. Rev. 39: 1-8.

Greig, R.A. and Gnaedinger, R. H. (1971). Occurrence of thiaminase in some common aquatic species of the United States and Canada. U.S.D.C.

Gunn, C. K. (1948a). Dominion of Canada, Dept. Agric. Exp. Fox Ranch, Summerside, P.E.I. Prog. Rep. 1936-1946.

Gunn, C. K. (1948). Summary of work at experimental fox ranch, P.E.I. for 1947. Amer, Fur Breeder 21: 26-28.

Hamilton, W. J. Jr. (1935). Notes on food of red foxes in New York and New England. J. Mammalogy 16(1): 16-21.

Hamilton, W. J. and Cook, D. B. (1944). The ecological relationship of red fox food in Eastern New York. Ecology 25(1): 91-104.

Hanson, K. B. (1935). Rickets in foxes - its prevention and cure. American Fur Breeder VIII (1):13.

Harman, R. D. (1954). Virginia fox and mink rancher. Personal communication.

Harris, L. E., Bassett, C.F., Llewellyn, L. M. and Loosli, J. K. (1951a). The protein requirement of growing foxes. J. Nutr. 43(1):167-180.

Harris, L. E., Bassett, C.F., Smith, S.E., and Yeoman, E.D. (1945). The calcium requirement of growing foxes. The Cornell Vet. 35(1): 9-22.

Harris, L.E., Bassett, C.F. and Wilke, C.F. (1951b). Effect of various levels of calcium and phosphorus and vitamin D intake on bone growth, I. Foxes. J. Nutr. 43: 153-165.

Harris, L. E. and Embee, N.D. (1963). Quantitative considerations of the effect of polyunsaturated fatty acids content of the diet upon requirements of vitamin E. Amer. J. Clin. Nutr. 13: 385.

Harris, L. E. and Loosli, J. K. (1949). The thiamine requirement of mature silver foxes. Cornell Vet. 39: 277-281.

Harris, L. E., Cabell, C. A.,Elvehjem, C. A., Loosli, J.K.and Schaefer, H. C.(1953). Nutrient Requirements For Domestic Animals Number VII. Nutrient Requirements For Foxes and Minks. National Research Council, Division of Biology and Agriculture. 2101 Constitution Avenue NW, Washiington, D.C.

Hatfield, D.M.(1939). Winter food habits of foxes in Minnesota. J. Mammalogy 20: 202-206.

Hayes, K.C., Carey, R.E. and Schmidt, S.Y. (l975). Retinal degeneration associated with taurine deficiency in the cat. Sci. 188: 949-951.

Heggset, O.S. (l998). Does extra acid in the feed at whelping time affect the incidence of metritis or litter size? Norsk Pelsdyrblad 72(3): 9-10.

Heinz Handbook of Nutrition (1959). McGraw-Hill Book Co. Inc. p. 31.

Heit, W. S. (1944). Food habits of red fox of the Maryland marshes. J. Mammalogy 25: 55-58.

Helgebostad, A. G. (1955). Experimental excess of vitamin A in fur animals. Nord Vet. Med. 7: 297-300.

Helgebostad, A. G. (1968). Vitamin B-1 - mancel hos mink og rev, Forbindelse med foring av sontorsk (Gadiculus thori). Norsk Pelsdyrblad 42: 619-622.

Helgebostad, A. G. (1971). Vitamin E, mangel syndromet hos pelsdyr. Norsk Vet. Tidsskr 83:11-17.

Helgebostad, A. G. (1976). Recent aspects of vitamin E deficiency in fur-bearing animals.1st Inter. Sci. Congress in Fur Animal Production. April 27-28, 1976. Helsinki, Finland.

Helgebostad, A. G, (1980). Vitamin C. I. Pelsdyrernaeringen. Norsk Pelsdyrblad 54: 161-162.

Helgebostad, A. G. and Bohler, N. (1949). Experimental rickets and tetany in the fox and dog. Vet. Rec. 61: 735-740.

Helgebostad, A. G. and Ender, E. (1955). Effect of feeding on pelt development in fox and mink. V. Marine fat the cause of discoloring in the pelt. Norsk Pelsdyrblad 8: 139-147.

Helgebostad, A. G. and Ender, E. (1973). Vitamin E and its function in the health and disease of fur-bearing animals. Acta Agr. Scand. Suppl. 19: 79-82.

Helgebostad, A. G. and Nordstogen, K. (1978). Hypervitaminosis D in fur Bearing animals. Nord. Vet. Med. 30:451-455.

Helgebostad, A. G., Svenkerud, R.R. and Ender, E. (1959). Experimental biotin deficiency in mink and foxes. Nord Vet. Med. 4: 141-144.

Hewson, R. and Kolb, H.H. (1975). The food habits of foxes (Vulpes vulpes) in Scottish forests. J. Zool. (London). 176: 287-292.

Hickman, K.C.D., Kaley, M.W. and Harris, P.L. (1944). Co-vitamin studies-III. The sparing equivalence of tocopherols and mode of action. J. Biol. Chem. 152: 321.

Hodson, A.Z. and Loosli, J. K. (1942). Experimental nicotinic acid deficiency in the adult silver fox. Vet. Med. 37(11): 470-473.

Hodson, A. Z. and Smith, S.E. (1942a). Thiamine deficiency and Chastek Paralysis in Foxes. The Cornell Vet. 32(3): 281-286.

Hodson, A.Z. and Smith, S.E. (1942b). Estimated maintenance energy requirements of foxes and mink. Fur Trade Journal of Canada 19(6): 12-15.

Hodson, A.Z. and Smith, S.E. (1945). Estimated maintenance energy requirements of foxes and mink. American Fur Breeder XVIII(4): 44-52.

Hoie, J. (1953). On soya cake and soymeal as food for fur animals. Norsk Pelsdryblad Sept. No. 11 - 8 pp.

Hoie, J. and Rimeslatten, H. (1950). Experiments in feeding proteins, fats and carbohydrates at different levels to silver and blue foxes. Institutt for Fjorfe og Pelsdyr. Norges Landbruk-hogskole, Melding nr. 4: 1-90.

Howell, R. F. (1957). Inquisitive guests attend nutrition conference. Amer.Fur Breeder 30(7): 16.

Ingo, R., Blomstedt, L., Tyopponen, J., Kangas, J. and Valtonen, M. (1989). Ghet hos silverrau. NJE Seminarium Nr. 1780. Stockholm, Sweden. 29 Sept-Oct 1.

Inman, W. R. (1941). Digestibility studies with foxes II. Digestibility of frozen beef, tripe, frozen lip meat, frozen beef hearts and frozen cow udders. Sci. Ag, 22: 33-39.

Inman, W. R. and Smith, G.E. (1941) Digestibility studies with foxes. I. Effect of the plane of nutrition upon the digestibility of meats. Scientific Agr. 22: 18-27.

Jaksic, F. M., Schlatter, R.D. and Yanez, J. L. (1980). Feeding ecology of Central Chilean foxes, Dusicyonclupaeus and Dusicyon griseous. J. Mammalogy 62(2): 254-260.

Jarl, E. (1944). Foods and feeding of fur animals. Landbrukshogsk Husdjurs-Forsoksanst Sartryck Forhandsmead No. 37, 1944. Vara Palsdjur 1944: 167-197.

Johnson, C..E. (1980). An unusual food source of the red fox (Vulpes vulpes). J. Zool. (London). 192 (4): 561.

Jones, W. G. (1960). Fishery Leaflet 501. Bureau of Comm. Fish. Fish and Wildlife Svc. Dept. Inter. P. 228.

Jongbloed, A. (1987). Phosphorus in the feeding of pigs. Academic dissertation, Instituut voor Veevedingsonderzoek (I.V. V. O.) Lelystad, The Netherlands.

Jorgensen, G. (1977). Vitamin D content in various species of fish and its influence on vitamin D content of mink feed. Dansk Pelsdyrblad 40:138-139.

Jorgensen, G. (2000). The use of soybean products in feeds for fur bearing animals. American Soybean Association, Pelzerstrasse 13, Hamburg, Germany.

Jorgensen, G. and Glem-Hansen, N. (1973). Fedtsyressam-mensaetningens indflydelse pa fedstoffernes fordojelighed Forsogslaboratoriet, Arbog.

Jorgensen, G., Hjarde, W. and Lieck, H. (1975). Intake, deposition and excretion of thiamine, riboflavin and pyridoxine in mink. 442 Beretning fra Forsogslaboratoriet: 37-54.

Kacmar, P., Samo, A. and Knezik, J. (1980). Chemical diagnostics of sodium chloride poisoning in thoroughbred fur-bearing animals, fox, coyote and in turkey and pheasants. Veterinarni Medicina 25(12): 733-738. Jorgensen, G, and ... . (1975).

Kaikusalo, A. (1971). On the breeding of the artic fox (Alopex lagopus) in NW Enonteko, Finnish Lapland. Suomen Riista 23: 7-16.

Kakela, R., Polonen, I., Miettinen, M. and Asikainen, J. (2001). Effects of different fat supplements on growth and hepatic lipids and fatty acids of mink. Acta Agr. Scand. 51(1): 217-223.

Kaneko, J. J., (1989). Carbohydrate metabolism and its diseases. In: Clinical Biochemistry of Domestic Animals. Kaneko, J.J.(ed) Academic Press, Inc. 44-61.

Kangas, J. (1976). Minerals in mink feed and factors affecting mineral balance. 1st Inter. Sci. Congr. In Fur Animal Prod. Helsinki, Finland.

Karpuleon, F. (1958). Food habits of Wisconsin fox. J. Mammalogy 39(4): 591-592.

Kechkaylo, W. (1991). Univ. California, Davis, California. Analyses of calculi provided to a Michigan fox rancher. Personal Communication.

Kittleson, M., University of California at Davis, California. Personal communication, 1987.

Kidder. H. (1988). Michigan fox rancher. Personal communication.

Kleckinym, L. (1940). Quoted in Perel'dik, N.S., Milovanov, L.V. and Erwin, A. T. (l972). Feeding fur animals. Translated from the Russian by the Agr. Res. Svc. U.S. Dept. Agr. And the Nat. Sci. Foundation, Washington, D.C. 344 pp.

Kleiber, M. (1961). The Fire of Life: An Introduction to Animal Energetics. Wiley, N.Y.

Koppang. A., Helgebostad, A., Armstrong, D. and Rimeslatten, H. (1981). Toxic and carcinogenic effects of dimethylnitrosamine (DMNA) in the blue fox (Alopex lagopus). Acta Vet. Scand. 22: 501-516.

Korschgen, L. J. (1959). Food habits of the red fox in Missouri. J. Wildlife Mgt. 23(2): 168-176.

Koskinen, N., Dahlman, T., Polonen, I.., Valaja, T. and Rekila, T. (2005a). Effects of two phase protein feeding of blue foxes during growing period, a field study. Proceedings of the 3rd Inter. Symposium. Physiological Basis for Increasing the Productivity of Mammals Introduced in Zooculture. Sept. 27-29, Petrozavodsk.

Koskinen, N., Dahlman, T., Polonen, I., Valaja,J. and Rekila, I. (2005b). Low protein and methionine in blue fox diet during growing-furring season, a field study. Proceed. Of the 3rd Inter. Symposium on Physiological Basis For Increasing the Productivity of Mammals Introduced in Zooculture. Petrozavodsk.

Kringstad. H. and Lunde, G. (1940). The importance of the vitamin B complex for silver foxes. Skand. Vet. 30: 1121-1128.

Kringstad, H. and Lunde, G. (1940a). Vitamin B-1 deficiency in silver foxes. Norsk Pelsdryblad 14: 5-8.

Kringstad, H. and Lunde, G. (1940b). Vitamin B-complexes betiding for solvreu. Skandinavisk Vet-Erinar-Tidskrift 30: 1121.

Kuusi, T. (1963). On occurrence of thiaminase in Baltic herring. Valtion Teknillinen Tutkimuslaitos. The State Institute for Technical Research, Helsinki, Finland. Tiedotus Sarja IV -Kema 52.

Lee, C.F. (1948). Thiaminase in fishery products. A Review. Comm. Fish Rev. U.S.D. I. F. W. S. Rep. No.202.

Lee, C. F., Nilson, H. W. and Clegg, W. (1955). Weight, range, proximate composition and thiaminase content of fish taken in shallow water trawling in Northern Gulf of Mexico. U.S.D. I.F. W.S. Rep. No. 396 Technical Note 31. Comm. Fish Rev. 17(3): 21.

Leoschke, W. L. (1956). Phosphoric acid in the mink's diet. Amer. Fur Breeder 29(8): 18.

Leoschke, W. L. (1957, 1982, 1983 and 1987-1989). National Fur Foods, Unpublished data.

Leoschke, W. L. (1959). The digestibility of animal fats and proteins by mink. Amer. J. Vet. Res. 20(79): 1986-1089.

Leoschke, W. L. (1960). Mink nutrition research at the Univ. Wisc. Research Bul. No. 222.

Leoschke, W. L. (1962). Studies on the wet-belly disease of the mink. Progress Reports - Mink Farmers Research Foundation.

Leoschke, W. L. (1963). Digestibility of cereal grain carbohydrates by the mink. Progress Reports - Mink Farmers Research Foundation.

Leoschke, W. L. (1969). American Fur Breeder Fur Farm Guide Book 42: 85-116.

Leoschke, W. L. (1987a). Carbohydrates in modern mink nutrition. Blue Book of Fur Farming 1987: 39-42.

Leoschke,m W. L. (1987b). Kohlenhydrate inder nerzfuterung (Importance of carbohydrates for fur productioin of mink. Der Deutsche Pelztierzucher 11:163-164 and l2: 177-179.

Leoschke, W. L. (1996). Phosphoric acid in modern mink and fox nutrition. 6th Inter. Sci. Congr. In Fur Animal Production. August 21-23, 1996. Warsaw, Poland. Anim. Prod. Review, Polish Society of Anim. Prod. Applied Science Reports 28.

Leoschke, W. L. and Chausow, D. (1987). Modern pellets versus ranch mix for foxes. 1988 Blue Book of Fur Farming, Nov. 1987. Pp. 72-74.

Leoschke, W. L., Lalor, R. J. and Elvehjem, C. A. (1953). The vitamin B-12 requirement of mink. J. Nutr. 541-545.

Leoschke, W. L., Zikria, E. and Elvehjem, C. A. (1952). Composition of urinary calculi of mink. Proc. Soc. For Exper. Biol. And Medicine 80:291-293.

Linhart, S. B. and Enders, R. K. (1964). Some effects of diethylstilbestrol on reproduction in captive red foxes. J. Wildl. Manage. 28:358.

Linscombe, G., Kinler, N. and Aulerich, R. J. (1982). Mink (Mustela vison) in J. A. Chapman and G.R. Feldhamer (Eds.) Wild Mammals of North America, Biology, Management, Economics. The John Hopkins University Press, Baltimore, Md. Pp. 629-649.

Loew, F.M. and Austin, R. J. (1975). Thiamine status of foxes with Chastek's Paralysis. Can. Vet. J. 16(2): 50-52.

Long, J. B. and Shaw, J. N. (1943). Chastek paralysis produced in Oregon mink and foxes by the feeding of fresh frozen smelt. North Am. Vet. 24:234.

Lund, G. and Kringstad, H. (1939). Requirements of the fox for anti-gray hair factor from vitamin Bx. Naturwissenschalften 27: 755.

MacDonald, D. (1992). The Velvet Claw - A Natural History of The Carnivores, BBC Books, London, 256 p.

Makela J. and Kiiskinen, T. (1975). Sojaforsok med blaravshvalper.Finsk Palslidskrift 9: 253-255.

Mamaeva, G. E. (1958). Influence of nutrition on the body measurements of silver foxes. Krolik i Zver. 3:25-30.

Mathiesen, E. (1939). Does the silver fox need vitamin C in its feed? Norsk Pelsdyrblad 13:456.

Mathiesen, E. (1942). Vitamin C in the diet of silver foxes. Skand. Veterinartid. Bact. Patol. Kott-och Mjolkhyg. P. 315.

McCarthy, B. , Travis, H. F., Krook, L. and Warner, R.G. (1966). Pantothenic acid deficiency in the mink. J. Nutr. 89: 392-396.

McGregor, A. E. (1942). Late fall and winter food of foxes in central Massachusetts. J. Wildlife Management 6(3): 221-224.

Melnick, D., Hochberg, M. and Oser, B. L. (1945). Physiological availability of the vitamins. II. The effect of dietary thiaminase in fish products. J. Nutr. 30(2): 81-88.

Mertin, D., Tocka, I. And Oravcova, E. (1991). Effect of zinc, Selenium on some morphological properties in silver foxes in period of fur maturity. Scientifur 15(4): 287-293.

Moe, R. O., Dille, L.L. and Bakken, M. (2000). Water requirements of farmed foxes. Scientifur 24(4) - Nutrition. Proc. Of the 7th Inter. Sci. Congr. In Fur Animal Production. Ed. B. D. Murphy and O. Lohi: 54-56.

Morgan, A.F. and Simms, H.D. (1940). Anti-grey hair vitamin deficiency in the silver fox. J. Nutr. 20(6): 627-635.

Mosshammer, H. (1985). Indiana fox rancher. Personal Communication.

Mulder, G. J. (1838). The chemistry of animal and vegetable physiology. Quoted in L. B. Mendel, Nutrition, The Chemistry of Life. Yale University Press, New London, Conn. 1923. p. 16.

Nelson, A. L. (1933). A preliminary report on the winter food of Virginia foxes. J. Mammalogy 14: 40-43.

Nelson, G. J. and Ackerman (1988). Absorption and transport of fat in mammals with emphasis on n-3 polyunsaturated fatty acids. Lipids 23: 1005-1014.

Neilands, J. B. (1947). Thiaminase in aquatic animals of Nova Scotia. J. Fish Res. Bd., Canada 7(2): 94.

Neseni, R. (1935). Einige anatomische datin von pelztieren prager tierarztle. Archieves 10: 211-216.

Neseni, R. and Piatkokski, B. (1958). Feed passage time in mink. Archtierenahe 8: 296.

Nenonen, N., Polonen, I., Rekila, T., Siirila, P. and Valaja, J. (2002) Dehulled and naked oats in mink and blue fox diets. Proceedings from NJF-Seminar No. 347. 6 pp.

Nenonen, N., Polonen, I., Rekila, T and Valaja, J. (2003). Influence of dietary vitamin A level on growth performance and fur quality of mink and blue fox.NJF-Seminar #354. Lillihammer, Norway. 10/8-10, 2003.

Nordfeldt, S., Melil, G. and Thelander, B. (1955). Digestion experiments with mink. Kungl. Lantbrukshogsk, Statens Husdjursforsok Sartryck Forhandsmedd. No. 114. Vara Palsdjur 1955 Nos. 5 and 7.

Osborne, C. A., Polzin, D.J., Kruger, M., Abdullahi, S. U., Leninger, J.R. and Griffth, D. P. (1986). Medical dissolution of canine struvite uroliths. Vet. Clin. of N. Amer. Small Animal Pract. 16(2): 349-374.

Osweiler, G. D., Carson, T. L., Buck, W. B. and Van Gelder, G. A. (1976). Clinical and Diagnostic Vet. Toxicology. 3rd Ed. Pp. 183-188.

Ott, G. L. and Coombes, A. I. (1941). Rickets in silver fox pups. Vet. Med. 36(4): 202-205.

Palmer, L.S. (1927). Dietetics and its relationship for fur. American Fox and Fur Farmer 7(2): 22-24.

Papageorgiou, N.K., Sepougaris, A., Christopoulou, G.G., Vlachos, C.G. and Petamidis, J.S. (1988). Food habits of the red fox in Greece. Acta Theriol. 33:313-324.

Pelletier, O., Keith, M.O. (1974). Bioavailability of synthetic and natural ascorbic acid. J. Am. Diabetic Assoc. 64(3): 271-275.

Penelaik, N. (1972).Feeding Fur Bearing Animals. Kormlenie Putnyh, Zverep.

Penelaik, D. (1975). Russian Handbook.

Perel' dik, N.S., Milovanov, L. V. and Erin, A.T. (1972). Feeding fur breeding animals. Translated from Russian by the Agricultural Research Service, U.S. Department of Ag. And National Sci. Foundation, Washington, D.C. 344 pp.

Perel'dik, N.S. and Perel'dik, D. N. (1980). Ferroanemin – preventing cotton fur. Krolikovodstov I. Zverodstva, Moskva 33:33.

Phillips, P.H. and Hart, E. B. (1935). The effect of organic dietary constituents upon chronic fluorine toxicosis in the rat. J. Biol. Chem. 109: 657658.

Picard, F. H. J. and Bakker, T. (1939). Physiological basis of poor quality pelts in fur animals in captivity. Deutsch. Pelztierzuchter 14:342. Iron and vitamin A content of the livers of silver foxes. Tijdschr Diergeneesk 66: 353-356.

Polonenm I. and Dahlman, T. (1988). Carbohydrates in diets for fur bearers. Finsk Palstidskrift 22(12): 487-489.

Polonen, I., Valaja, J., Jalava, T., Perttila.S., Sauna-Aho, R. and Kariloto, S. (2000). Effect of dietary folic acid supplementation on formate metabolism in blue foxes (Alopex lagopus) . Scientifur 24(4). IV-A Nutrition. Proc. Of the 7th Inter. Sci. Congr. In Fur Animal Productionm ed. B. D. Murphy and O. Lohi: 32-35.

Pozdnjakov, E.V. (1955). Metabolism in young blue foxes. Karakul. Zver. 8(6): 50-53.

Quist, A. (1964). Natriumbsulfite konserverat foder. Dansk Pelsdyravl 27: 412-413.

Rapoport, O. L. (1961a). Influence of fat on reproduction capacity of blue foxes. Krolik. Zver. No. 2:16.

Rapoport, O. L. (1961b). Effect of increased amounts of fish oil. Krolik. Zver. No. 7 16: 20.

Rapoport, O. L. (1961c). Addition of fats to rations for blue foxes. Krolik. Zver. No. 12: 19.

Raymond, A. (1980). Swedish mink and fox rancher. Personal communication.

Richards, D. F. (1977). Observations on the diet of the red fox (Vulpes vulpes) in South Devon. J. Zool. (London). 183: 495-504.

Rimeslatten,H. (1951). Forsok med koking av kollhydraterk raft foret til solrevblarev-og minkhvalper. Norsk Pelsdyrblad, 25: 329-332.

Rimeslatten, H. (1957). Eksperimentell ribflavinmangel hos blarevhavlper. Norsk Pelsdyrblad 31: 84-96, 123-135.

Rimeslatten, H. (1958). Experimental riboflavin deficiency in blue fox cubs. Effect of riboflavin on pelt color and pelt quality. Deutsch Pelztierzachter 32: 142-145.

Rimeslatten, H. (1959). Trials with vitamins, animal liver and trace elements for silver fox and mink. Dansk Pelsdyravl 22: 273-275.

Rimeslatten, H. (1968). Minkens og revens til a lagre vitamin A i leveren under forskellige fodringforhold. Norsk Pelsdyrblad 42: 542-546.

Rimeslatten, H. (1974). Soyamjol som for til pelsdyr. Norsk Pelsdyrblad 48: 219-223.

Rimeslatten, H., (1976a). Experiments in feeding different levels of protein, fat and carbohydrates to blue foxes. The lst Inter. Sci. Congr. In Fur Animal Production, Helsinki, Finland. April 27-29, 1976. 29 pp.

Rimeslatten, H. (1976b). Experiments in feeding different levels of protein, fat and carbohydrates to blue foxes. Scientifur 1: 28-35.

Rimeslatten, H. (1978). Solvrevens emergibehov og forets sammensaetning. The Scandinavian Assoc. of Ag. Sci. Meeting in Helsingor, Denmark, 1978. Pp. 30.

Rimeslatten, H. (1982). Unpublished data . Provided in Nutrient Requirements Of Domestic Animals, Number 7. Nutrient Requirements of Mink and Foxes. Board on Agriculture and Renewable Resources. National Research Council, National Academy Press, Washington, D.C. 1982.p. 36.

Rimeslatten, H. and Aam, A. (1962). Forsoek med torrfiskmjol til solrev, blarrev or mink. Norsk Pelsdyrblad 36: 392-396.

Robertson, P. A, ad Whelan, J. (1987). The food of the red fox (Vulpes vulpes) in Co. Kildare, Ireland. J. Zool. London 213: 740-743.

Rogers, Q. R., Baker, D. H., Hayes, K.C., Kendall, P.J. and Morris, J. C. (1986). Nutrient Requirements of Cats, Revised edition (1986). Nutrient Requirements of Domestic Animals. National Academy Press, Washington. D.C.

Rouvinen, K. I. (1987). Olika fetter I palsdjursfoder. NJF-seminarium nr. 128. Mote om paladjursproduktion, Tromso, Norway 28-30.

Rouvinen, K. I. (1991). Dietary effects of omega-3 polyunsaturated fatty acids on body fat composition and health status of farm-raised blue and silver foxes. Acta Agr. Scand. 41(4): 401-414.

Rouvinen, K. I. (1992). Effects of fish fat feeding on body fat composition of foxes. Norwegian J. of Ag. Sci. Supplement #9, 1992. Progress in fur animal science. Proceedings from the 5th Inter. Sci. Congr. In Fur Animal Production. :249-253.

Rouvinen, K. I., Anderson, D.M. and Alward, S. R. (1997). Effects of high dietary levels of silver hake and Atlantic herring on growing-furring performance and blood clinical-chemistry of mink (Mustela vison). Can. J. Anim. Sc. 77: 509-517.

Rouvinen, K. I., Archbold, S., Laffin, S. and Harri, M. (1999). Long-term effects of tryptophan on behavior response and growing–furring performance of silver fox (Vulpes vulpes). Applied Animal Behavior Science 63: 65-77.

Rouvinen, K. I., Inkinen, R. and Niemela, P. (1991). Effects of slaughter house offal and fish mixture based diets on production performance of blue and silver foxes. Acta Agr. Scand. 41(4): 387-399.

Rouvinen, K. I. And Kiiskinen, T. (1988). Digestibility of different fats and fatty acids in blue fox (Alopex lagopus). Biology, Pathology and Genetics of Fur Bearing Animals. Ed. B. D. Murphy and B.B. Hunter. Proc. 4th Inter. Sci. Congr. In Fur Animal Production. Rexdale, Ontario, Canada, August 21-24, l988: 336-343.

Rouvinen, K. I. And Kiiskinen, T. (1989). Influence of dietary fat on the body fat composition of mink (Mustela vison) and blue fox (Alopex lagopus). Acta Agr. Scand. 39: 279-288.

Rouvinen, K. I. And Kiiskinen, T. (1991). High dietary ash content decreases fat digestibility in the mnk. Acta Agr. Scand. 41(4): 375-386.

Rouvinen, K.I., Kiiskinen, T. and Makela, J. (1988). Digestibility of differentfats and fatty acids in the blue fox (Alupex lagopus). Acta Agric. Scand. 38: 405-412.

Rouvinen, K. I. and Laine, T. (1991). Acute toxic effects of ethoxyquinine in the blue fox. Scientifur 15(4): 328.

Rouvinen, K. I., Makela, J., Kisskinen, T. and Nummela, S. (1992). Accumulation of dietary fish fatty acids in the body fat reserves of some carnivorous fur-bearing animals. Agrc. Sci. Finl. 1-7.

Rouvinwn, K.I. And Niemela, P. (1992). Long-term effects of dietary fish fatty acids on the breeding performance of blue foxes. Scientifur 16(2): 143-151.

Rouvinen, K. I., Niemala, P. and Kiiskinen, T. (1989). Influence of dietary fat source on growth and fur quality of mink and fox. Acta Agri. Scand. 39: 269-278.

Rusanen, M. and Valtonen, M. (1991). Blue fox milk composition. Scientifur 15(4):327

Samuel, D. E. and Nelson, B.B. (1982). Foxes (Vulpes vulpes) and allies. In J. A. Chapman and G. A. Feldhamer (Eds.). Wild Mammals of North America, Biology, Management, Economics. The John Hopkins University Press, Baltimore, MD, p. 475-490.

Schaefer, A. E., Whitehair, C.K. and Elvehjem, C. A. (1947a). The importance of riboflavin, panto-thenic acid, niacin and pyridoxine in the nutrition of foxes. J. Nutr. 34(2): 131-139.

Schaefer, A.E., Whitehair, C.K. and Elvehjem, C. A. (1947b). Purified rations and the requirement of folic acid for foxes. Archives of Biochemistry 12(3): 349-357.

Schaefer, A. E., Whitehair, C.K. and Elvehjem, C. A. (1948). Unidentified factors essential for growth and hemoglobin production in foxes. J. Nutr. 35(2): 147-156.

Schaefer, A.E., Tove, S. B., Whitehair, C. K. and Elvehjem, C. A. (1947). Use of foxes and minks for studying new B vitamins. Tirage a'part de "Zeitschrift fur Vitaminforschung" tome XIX - Cahier 1/2 - 1947. Ed. Hans Huber, Berne.

Scheuler, R. L. (1951). Red fox food habits in a wilderness area. J. Mammalogy 32(4): 462-464.

Schoop, G. (1939). Inadvisability of giving calcium supplements to foxes and mink. Deutsch. Pelztierzucher 14: 332-336.

Schultz, T. D. and Ferguson, J. H. (1974). The fatty acid composition of subcutaneous, ometal and inguinal adipose tissue in the Artic fox (Alopex lag opus innuitus). Comp. Biochem. Physiol. 49B: 65-69.

Sibbald, I.R., Sinclair, D. G., Evans, E.V. and Smith, D. L.T. (1962). The rate of passage of feed through the digestive tract of the mink. Canadian J. of Biochem. And Physiology 40:1391-1394.

Skrede, A., Krogdahl, A. and Austreng, E. (1980) Digestibility of amino acids in raw fish flesh and meat-and-bone meal for the chicken, fox, mink and rainbow trout. Zeitschrift Fir Tierphysiologic Tierphysiol. Tierenahrg u Futtermittelkde Band 43: 92-101.

Skrede, A. (1981). Varierande feitt:karbohydrate-tilhaove i for til mink. I. Verknad pa reproduksjon vekst leveevne og kjemisk innhald I kroppen hja kvelpoar. Meld. Norg. LandbrHogsk. No. 16, 60: 20 pp.

Skrede, A. and Ahlstrom, O. (2001). Meat-and-bone meal as feed for blue foxes and mink in the reproduction period. NJF-Seminar No. 331. 6 pp.

Skrede, A. and Ahlstrom, O. (2002). Bacterial protein produced on natural gas: A new potential feed ingredient for dogs evaluated using the blue fox as a model. J. Nutr. 132: 1668S-1669S.

Skrede, A. and Ahlstrom, O. (2004). Bacterial protein produced on natural gas as a protein source in dry diets for the growing-furring blue fox. Scientifur 28(3): 188-192.

Skrede, A. and Gulbrandsen, K.E. (1985). Fettkilder I. Pelletert torrfor till mink or blarev. JF mode vedr, pelsdyrproduktion. NJ.Seminarium Nr. 85, Aalborg, Denmark 3-5, 9.

Skrede, A., Krogdahl, A. and Austreng, E. (1980). Digestibility of amino acids in raw fish flesh and meat-and-bone meal for chicken, fox, mink and rainbow trout. Zeitschrift Fir Tierphysiologic, Tierenahrung und Futtmillekunde. Band 43(2): 92-101.

Smith, G. E. (1927). Progress Reports - Results of Experiments 1926-1927. Dominion Experimental Fox Ranch, Summerside, Prince Edward Island, Canada.

Smith, H.J. (1932). Foxes thrive on exclusive Purina Chows. Amer. Fur Breeder V(1): 14-15.

Smith, G.E., (1935). Experimental Fox Ranch Reports - 1931, 1932, 1933 and 1934. Dominion Experimental Fox Ranch, Summerside, Prince Edward Island, Canada.

Smith, S.E. (1941a). Vitamin A deficiency in silver foxes. Am. Fur Breeder 14(3): 10-12.

Smith, S.E. (1941b). Blood glucose, plasma inorganic phosphorus, plasma calcium, hematocrit and bone ash values of normal minks (Mustela vison) and foxes (Vulpes fulva). The Cornell Veter. 51: 56-62.

Smith, S.E. (1942a). The digestibility of some higher protein feeds by fox. Archives of Biochem. 1: 263-267.

Smith, S.E. (1942b). The minimum vitamin A requirements of the fox. J. Nutrition 24(2): 97-109.

Smith, S.E. and Barnes, L.L. (1941). Experimental rickets in silver foxes (Vulpes fulva) and minks (Mustela vison). Unpublished data. Cited by Harris, L. E., Basset, C. F. and Wilkie, C (1951). Soc. Exp. Bio. 1942 60: 268-269.

Spector, W. S. (1956). Handbook of Biological Data. Division of Biology and Agr., Nat. Academy of Sci., National Research Council, Saunders. P. 50.

Steger, H. and Piatkowski, B. (1959). Nutrient content and digestibility of silkworm pupa meal in studies on silver foxes and mink. Arch. Tierernahrung 9:463.

Stevens, C. E. (1977). Comparative Physiology of the Digestive Systems. In Swenson, M.J. (ed.) Dukes' Physiology of Domestic Animals. 9th ed. Cornell University Press, Ithaca, New York. USA 14850: 216-232.

Stoddart, D.M. (1974). Earthworms in the diet of red fox (Vulpes vulpes). J. Zool. (London) 173: 251-275.

Stout, F. M., Oldfield, J. E. and Adair, J. (1963). A secondary induced thiamine deficiency in mink. Nature 197: No. 4869: 810-811.

Stout, F. M., Baily, D.E., Adair, J. and Oldfield, J.E. (1968). Iron - a chromomeric nutrient. J. Anim. Sci. 27(4): 1157.

Stowe, H.D. and Whitehair, C. K. (1963). Gross and microscopic pathology of tocopherol deficient mink. J. Nutr. 81: 287-300.

Szymeczko, R., Bieguszewski, H. and Burlikowska, K. (1996).The influence of dietary fibre on nutrient digestibility in polar foxes. Animal Production Review, Polish Society of Animal Production. Applied Science Reports 28. Progress in Fur Animal Science, Nutrition, Patology and Disease. Proceedings from the 6th Inter. Sci. Congr. In Fur Animal Production. August 21-23, 1996. Warsaw, Poland. Ed. A. Frindt and M. Brzozowski.

Szymeczko, R., Jorgensen, G., Bieguszewski, H. and Borsting, C. (1992). The effect of protein source on digestive passage and nutrient digestibility in polar foxes. Norwegian J. of Ag. Sc. Supplement No. 9, 1992, Progress in Fur Animal Science, Proceedings From the 5th Inter. Sc. Congr. In Fur Animal Production. August13-16, Oslo, Norway.

Szymeczko, R. and Skrede. A. (1990). Protein digestion in mink. Acta Agr. Scand. 40(2):, 189-200.

Szymeczko, R. and Skrede, A. (1991). Protein digestion in fistulated polar foxes. Scientifur 15(3): 227-232.

Tallas, P.G. and White, R.G. (1988). Glucose turnover and defense of blood glucose levels in Artic fox (Alopex lagopus). Comp. Biochem. Physiology 91A (3): 493-498.

Taylor, P. G., Marinez-Torres, C., Romano, E. L. and Layrisse, M.(1986). The effects of cysteine containing peptides released during meat digestion on iron absorption in humans. Am. J. Clin. Nutr. 43: 68-71.

Tauson, A-H., Olafsson, B.L., Elnif, J., Treuthardt, J.and Ahlstrom, O. (1992). Minkens och Ravens Mineralforsorjning. NJF-Utredning/Rapport Nr. 79, Copenhagen, Denmark 1992. 104 p.

Titova, M. I. (1950). Nutritive adequacy of dried skimmed curds. Karakul. Zver. 3, No. 3: 75.

Titova, M. I. (1959). Digestibility of whale meat for blue foxes. Krolik. Zver. No. 5:18.

Tolonen, A. (1982). The food of the mink (Mustela vison) in North Eastern Finish Lapland in 1967-1976. Suomen Riista 29: 61-65.

Tove, S. B., Schaefer, A.E. and Elvehjem, C. A. (1949). Folic acid studies in the mink and fox. J. Nutr. 38: 469-478.

Travis, H.F., Evans, E.V., Jorgensen, Aulerich, R. J., Leoschke, W. L. and Oldfield, J.E. (1982). Nutrient Requirements of Domestic Animals Number 7. Nutrient Requirements of Mink and Foxes. Second revised ed. 1982. National Academy Press, 2101 Constitution Avenue, Washington, D.C., 20418.

Travis, H.J. and Pilbean, J.E. (1978). Effects of storage on the vitamin E and oxidative rancidity levels of feeds. Fur Rancher 58(2): 10-11.

Tyopponen, H., Berg,H. and Valtonen, M. (1987). Effects of dietary supplement of methionine and lysine on blood parameters and fur quality in blue fox during low-protein. J. Agr. Sci. Finland: 59:355-360.

USDA, National Nutrient Data Base for Standard Reference, Release 20(2007).

Utne, F. (1974). B-vitamin i minkfodringen. Post-graduate course in mink nutrition. P. 23.

Van Campen, D.R., (1969). Copper interference with the intestinal absorptionof zinc-65 by rats. J. Nutr. 97: 164-168.

Van Campen, D. R. and Scaifi, P.U. (1967). Zinc interference with copper absorption in rats. J. Nutr. 91: 473-476.

Varnish, S.A. and Carpenter, K. J. (1975). Mechanisms of heat damage in protein, the nutritional values of heat damaged and propionylated proteins as sources of lysine, methionine and tryptophan. Br. J. Nutr. 34:325-328.

Valaja, J., Polonen, I., Jalava, T., Perttila, S. and Niemela, P. (2000). Effectsof dietary mineral content on mineral metabolism and performance of growing blue foxes. Scientifur 24(4). IV-A Nutrition. Proc. Of the 7th Inter. Sci. Congress in Fur Animal Production. Ed. B. D. Murphy and O. Lohi.

Valaja, J., Polonen, I., Nenonen, N. and Jalava, T. (2003). Diets containing high amounts of barley for growing blue foxes. NJF-Seminar nr. 354, Lillehammer, Norway. Oct. 8-10, 2003.

Valaja, J., Polonen, I., Rekila, T., Nenonen, N. and Jalava, T. (2002). Calcium and phosphorus nutrition of blue foxes. Proceedings from NJF-Seminar No. 347. Nordiske Jordbrugs Forskeres Forening, Vokatti, Finland, 2-4, October, 6 pp.

Valaja, J., Polonen, I., Valkonen, E. and Jalava, T. (2004). Effects of lactic acid bacteria on the nutritive value of barley for growing fox. Scientifur 28(3)/ VIII Inter. Sci. Congr. In Fur Animal Production. DeRuwrenberg's Hertogenbosch, The Netherlands 15-18, Sept. 2004: 100-102.

Van Campen and Scaife (1967).

Vermeulen, C.W., Ragens, H.D., Grove, W. J. and Goetz, R. (1951) Experimental urolithasis III. Prevention and dissolution of calculi by alteration of urinary pH. J. Urology 66(1): 1-5.

Virtanen Blue Fox Farm, British Columbia, Canada (1982). Personal communication.

Watson, A. (1976). Food remaines in the dropping of foxes (Vulpes vulpes) in the Cairngorms. J. Zool. (London). 180: 495-523.

Westlake, R.L. (1989). DVM, Detroit Lakes, Minn.Vet. Clinic. Personal communication.

White, M. B., Egan, L.A., and Anderson, D.M. (1992). The effects of dietary acidifiers in diets of mature ranched foxes with a history of chronic urolithiasis. Norwegian J. of Agri. Sci. Supplement #9. 1992. Progress in Fur Animal Science. Proc. Of the 5th Inter. Sci. Congress in Fur Animal Production. Oslo, Norway : 326-331.

Wilson, H. C. (1983). Wilson and Partners, Veterinary Surgeons, 136 Bonnygate, Cupar, Fife KY15 4LF, Scotland. Personal communication.

Wolf (1942).

Wood, A.J. (1956). Time of passage of food in mink. The Black Fox Magazine and Modern Mink Breeder 39(9): 12-13.

Young, E. G. and Grant, G.A. (1931). The composition of vixen milk. J. of Biol. Chem. 93: 805-810.

Yu, T.C. and Sinnhuber, R. O. (1967). Developmemnt of a fat quality test for application to mink feeds. Projects of the Mink Farmers Research Foundation.

Zotova, V. S. (l968). Trace elements in tuhe ration of silver black foxes. Beloruss, Nauch-Issled. Inst. Zhivolovod 1968: 235-237.

WLL Fox Nutrition Book - Index

Index